Springer Proceedings in Energy

More information about this series at http://www.springer.com/series/13370

Laltu Chandra · Ambesh Dixit
Editors

Concentrated Solar Thermal Energy Technologies

Recent Trends and Applications

Springer

Editors
Laltu Chandra
Department of Mechanical Engineering and Center for Solar Energy
Indian Institute of Technology Jodhpur
Jodhpur, Rajasthan
India

Ambesh Dixit
Department of Physics and Center for Solar Energy
Indian Institute of Technology Jodhpur
Jodhpur, Rajasthan
India

ISSN 2352-2534 ISSN 2352-2542 (electronic)
Springer Proceedings in Energy
ISBN 978-981-10-4575-2 ISBN 978-981-10-4576-9 (eBook)
https://doi.org/10.1007/978-981-10-4576-9

Library of Congress Control Number: 2017945265

Printed on acid-free paper

This Springer imprint is published by Springer Nature
The registered company is Springer Nature Singapore Pte Ltd.
The registered company address is: 152 Beach Road, #21-01/04 Gateway East, Singapore 189721, Singapore

Preface

Concentrated solar thermal (CST) technologies offer versatility, in terms of utilization over other renewable energy technologies, especially, in the Indian scenario. To explore and report the recent developments in this field, the National Conference on Solar Thermal Energy Technologies (NCSTET) is organized during February 26–28, 2016 at Indian Institute of Technology Jodhpur, India. More than hundred delegates from India have actively participated in this event and shared their findings. Also, some renowned speakers were invited for the event to share their experiences on the design and development of CST systems and sub-systems with the audience. This volume provides the current trends in CST research and technology, associated issues, and challenges, especially in the Indian context.

Jodhpur, India
January 2017

Laltu Chandra
Ambesh Dixit

Conference Organizing Committee

Patron

Professor C.V.R. Murty, Director, Indian Institute of Technology Jodhpur, India

Scientific Advisory Committee

Shri A. Bhatewara, Clique Solar, Mumbai, India
Professor A. Subrahmanyam, Indian Institute of Technology Madras, India
Professor B. Bandyopadhyay, Retired Director, Solar Energy Centre, Gurgaon, IIEST Shibpur, India
Shri D. Gadhia, Gadhia Solar Energy Systems, Gujarat, India
Dr. E.Z. Moya, CIEMAT-Plataforma Solar de Almeria, Spain
Professor H.P. Garg, Indian Institute of Technology Delhi, India
Dr. H. Barshilia, National Aerospace Laboratory, Bangalore, India
Shri J.S. Solanki, General Manager, Godawari Green Energy Limited, Rajasthan, India
Professor K.L. Chopra, Former Director, Indian Institute of Technology Kharagpur West Bengal, India
Dr. M. Romero, Deputy Director, IMDEA Energy Institute, Spain
Professor N.K. Bansal, Retired, Indian Institute of Technology Delhi, India
Dr. N.M. Nahar, Retired Principal Scientist, CAZRI, Jodhpur, India
Professor P. Muthukumar, Indian Institute of Technology Guwahati, India
Dr. P.K. Vijayan, Bhabha Atomic Research Centre, Mumbai, India
Professor P.S. Ghoshdastidar, Indian Institute of Technology Kanpur, India
Professor R. Shekhar, Indian Institute of Technology Kanpur, India
Professor R. Banerjee, Indian Institute of Technology Bombay, India
Professor S.B. Kedare, Indian Institute of Technology Bombay, India

Shri S. Biswas, Director, Bharat Heavy Electricals Limited, Engineering R&D, Delhi, India
Professor S.S. Murthy, Emeritus Professor, Indian Institute of Science, Bangalore, India
Professor S. Singh, Indian Institute of Technology Bombay, India
Professor Y. Goswami, University of South Florida, USA

Convenor

Ambesh Dixit, Indian Institute of Technology Jodhpur, India
Laltu Chandra, Indian Institute of Technology Jodhpur, India

Local Organizing Committee

Barun Pratiher, Indian Institute of Technology Jodhpur, India
Deepak M. Fulwani, Indian Institute of Technology Jodhpur, India
Prodyut R. Chakraborty, Indian Institute of Technology Jodhpur, India
Rahul Chhibber, Indian Institute of Technology Jodhpur, India
V. Narayanan, Indian Institute of Technology Jodhpur, India

List of Reviewers

Dr. Antoinne Bittar, Managing Director, Front-End Solar Technologies Ltd., Germany and New Zealand
Dr. A.G. Shaik, IIT Jodhpur, India
Professor B. Bandyopadhyay, Ex-Director Solar Energy Center Gurgaon, Senior Advisor USAID
Professor B. Ravindra, IIT Jodhpur, India
Dr. E.Z. Moya, CIEMAT-Plataforma Solar de Almeria, Spain
Dr. Jim Hinkley, CSIRO Australia
Dr. M. Romero, Deputy Director, IMDEA Energy Institute, Spain
Professor N.K. Bansal, Retd. IIT Delhi, India
Dr. N.M. Nahar, Retd. CAZRI, Jodhpur, India
Dr. P. Chakraborty, IIT Jodhpur, India
Dr. P.K. Vijayan, Retd. BARC Mumbai, India
Professor P. Muthukumar, IIT Guwahati, India
Professor P.S. Ghoshdastidar, IIT Kanpur, India
Professor R. Shekhar, IIT Kanpur, India
Professor S.A. Khaprade, IIT Bombay, India
Professor S. Bandyopadhyay, IIT Bombay, India
Professor S.B. Kedare, IIT Bombay, India
Professor S.C. Mullick, IIT Delhi, India
Professor S.P. Revankar, Purdue University, USA
Dr. S. Shaktivel, ARCI, Hyderabad, India
Professor S. Singh, IIT Bombay, India
Professor Tolga Pirasaci, University of South Florida, USA
Professor T. Sundararajan, IIT Madras, India
Professor Y.D. Goswami, University of South Florida, USA

Contents

Part VI Solar Cooling

About the Editors

Dr. Laltu Chandra (Ph.D., 2005, University of Karlsruhe, Germany) has experience in experimental and numerical heat transfer and fluid mechanics. His interest includes nuclear and concentrated solar thermal sub-systems design, development, and evaluation. Recently, he has developed open volumetric air receiver based solar convective furnace system for metals processing. He has authored several research publications in international journals, as book chapters, and in conference proceedings.

Dr. Ambesh Dixit (Ph.D., 2010, Wayne State University, MI, USA) has experience in computational and experimental condensed matter physics with a special emphasis on design and development of materials for energy applications. He is currently working on the development of materials for energy conversion and storage. He has authored several research publications in international journals and conference proceedings.

Part I
Invited Papers

Innovative Uses of Solar Thermal Technology

Rajiv Shekhar

Abstract Solar thermal technology (STT) is versatile as it can be used to simultaneously generate electricity, process heat, and cooling. Initially, the emphasis of STT was on, and rightly so, large-scale electric power generation. A significant proportion of worldwide energy consumption ($\sim 60\%$) is in the form of heat for domestic and industrial use, for which STT can play a very important role (De Decker in The bright future of solar thermal powered factories, 2011) [1]. Substantial progress has been made in supplying solar thermal heat for domestic uses through flat plate collectors and concentrators. Use of solar thermal heat for vapour absorption cooling has also made reasonable progress. Some examples do exist where solar process heat (80–650 °C) and solar steam is being used in industries ranging from dairy to automobiles to textiles to pharmaceuticals (De Decker in The bright future of solar thermal powered factories, 2011) [1]. However, industrial applications of high-temperature solar thermal process heat are still in a state of infancy. In this paper, we will focus on some of the innovative applications of STT.

1 Metals Processing Operations

Metals processing operations are perfect candidates for STT as they use enormous amounts of fossil fuel directly or indirectly as electricity. There are several examples of the use of STT as a source of heat for the production of metals, such as zinc, by the carbothermic reduction of their oxides [2]. Here solar absorption furnaces (SAF) have been used. In SAF, concentrated solar radiation from heliostats is directed on to the reactants through a cavity in the reactor. The operating temperature could be as high as 1300–1500 °C. Details of SAF can be found in [2]. SAFs are still at an experimental stage.

R. Shekhar (✉)
Department of Materials Science and Engineering, Indian Institute of Technology Kanpur, Kanpur 208016, India
e-mail: vidtan@iitk.ac.in

L. Chandra and A. Dixit (eds.), *Concentrated Solar Thermal Energy Technologies*, Springer Proceedings in Energy, https://doi.org/10.1007/978-981-10-4576-9_1

One area where STT could be integrated with metals processing operations is metallurgical furnaces. One such example is an aluminium soaking furnace (ASF) where aluminium slabs are preheated to 600 °C for hot rolling. A schematic diagram of a 10 slab industrial ASF is shown in Fig. 1.

An ASF operates in a batch mode. The slabs are loaded on to a grid in the mid-section of the furnace. Electrically heated air is circulated to heat the slabs. The strategy for integrating the ASF with STT is shown in Fig. 2.

Hot air from an open volumetric air receiver (OVAR) based solar tower technology is used to supply hot air to the retrofitted ASF through a thermal energy storage system [2]. An existing ASF can be easily retrofitted by providing entry ports for hot air in the side sections and exit port(s) in the mid-section. The cooler exit air is sent back to the OVAR. Details of the OVAR design are given in [3]. During the day, the air temperature from the OVAR will vary in consonance with solar irradiance. Hence, a thermal energy storage system is used as a means to ensure that air at a constant temperature is continuously supplied to the ASF. A stand-by thermal energy storage system is also shown in Fig. 2. It would be best to operate the retrofitted ASF in a hybrid mode: solar heating would be the pre-dominant source during the day, while electrical heating gradually taking over with a decrease in solar irradiance. Some details of the retrofitted ASF design are given in [2].

Use of STT in aluminium smelting is another possibility. Aluminium smelters are electricity guzzlers. In India, most of the smelters have captive coal-based thermal power plants. STT can be hybridized with the existing thermal power plants. The STT cost, a major impediment, would be reduced by nearly 40% as it will share the existing plant's generator.

Solar thermal heat has also been used in electrolytic refining of copper at the Gaby mine in Chile since June 2013 [4]. A 27.5 MW_{th} solar thermal plant supplies 83% of the heat required to maintain the electrolyte temperature at 48 °C, and in the process significantly reducing fossil fuel consumption.

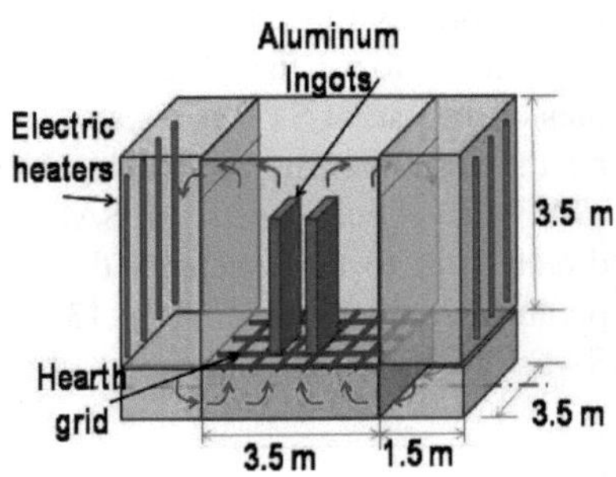

Fig. 1 A schematic diagram of a 10 slab industrial aluminium soaking furnace [2]. *Arrows* denote the flow of air in the furnace

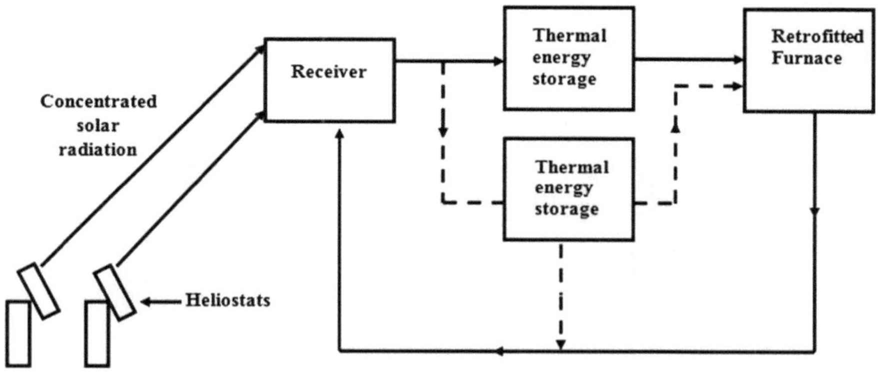

Fig. 2 Strategy for integrating solar thermal technology with a retrofitted ASF [2]

2 Solar Fuel Production

Solar fuel is a higher energy fuel which is produced from a lower energy resource such as natural gas or petroleum coke (PC) using solar energy. The additional energy comes from the "entrapment" of solar energy to form synthetic gas (syngas), which is a mixture of hydrogen and carbon monoxide. Both natural gas and petroleum coke undergo steam gasification reactions: (i) $CH_4 + H_2O = 3H_2 + CO$ and (ii) $C + H_2O = CO + H_2$.

Figure 3 shows an experimental petroleum coke reactor. It is in the form of a cylindrical cavity, 201 mm long, with an inside diameter of 120 mm [5]. The front opening of the reactor is covered by a quartz window to insulate the reactor from the surroundings. The reaction takes place in the region beyond the aperture. 142 μm PC particles and steam, both at 150 °C, were injected circumferentially at mass flow rates between 31–76 and 61–150 mg/s respectively. Heat for the endothermic reaction between PC and steam is provided by concentrated solar radiation, passing through the 50 mm aperture. The intensity of solar radiation varied from 1680 to 3360 kW/m^2 of aperture area. PC particles and steam formed a vortex with a residence time varying between 0.69 and 1.55 s. The reactor temperature varied from 1023 to 1545 °C. Syngas, along with the unreacted PC particles and steam, exited through a pipe at the back end of the reactor. In a single pass, 87% of the PC was converted into syngas. It may be pointed out that the low temperature steam required for gasification can also be produced by STT. Further details of the reactor are given in [5].

High temperature (750–850 °C) steam reforming of natural gas is a well-known industrial process to produce syngas. However, high temperatures and the non-continuous nature of solar irradiance had stalled the use of STT, which with the present technology can only give temperatures up to 550 °C. In addition to the steam-natural gas reforming (SNGR) reaction, the second reaction is the water-shift (WS) reaction, $CO + H_2O = CO_2 + H_2$, which takes place in the temperature range

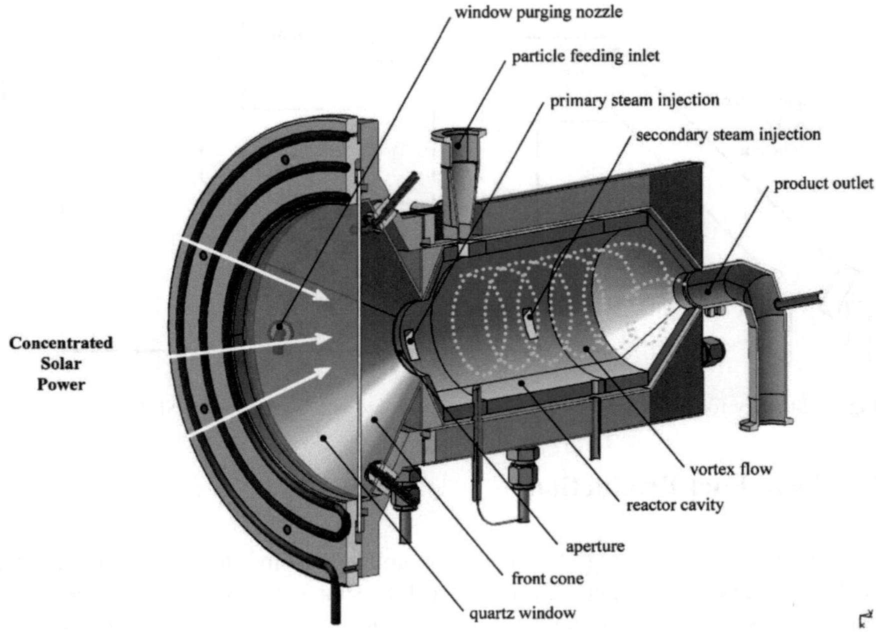

Fig. 3 Schematic diagram of a reactor used for the steam gasification of petroleum coke [5]

of 350–400 °C. Both reactions require catalysts for fast kinetics. Hence the trick was to somehow have appreciable natural gas reforming kinetics at a temperature less than 550 °C. Scientists at CSIRO used a membrane reactor [6], which in essence is a cylinder filled with catalysts, surrounded by a supported Pd membrane. Natural gas and steam are injected from one end of the reactor, which is heated through a counter-current hot air heat exchanger. The H_2 produced by the two reactions is absorbed by the Pd membrane, thereby driving both the SNGR and WS reactions to the right, and in the process significantly increasing their kinetics. CSIRO also developed two bimetallic CeO_2–La_2O_3-supported Cu–Cu-based catalysts, which showed greater activity than the Ni-based catalyst (for natural gas reforming) and Fe–Cr based catalyst (for water-shift reaction) 550 °C and 1 atm. Laboratory tests achieved a CH_4 conversion of 95%. Details of prototype reformer testing and integration with solar thermal energy are given in [6].

3 Enhanced Oil Recovery (EOR)

Global recoverable reserves of heavy and viscous crude oil amount to nearly 430 billion barrels. High viscosity, between 0.1 and 10 Pa s, is a major impediment to the recovery of these oils by conventional methods. Increasing crude oil temperature significantly reduces its viscosity making it easier to pump. Steam injection is the most common method of heating crude oil. In EOR, steam is pumped down the well into the oil reservoir. Oil is heated and pressurized, forcing it out of the well. In addition, in oil-bearing strata, where steam cannot penetrate by capillarity, the temperature of oil increases due to the conductive heating of the strata [7]. Either a part of the recovered crude oil or natural gas is the most important heat source for the generation of steam. However, both processes release greenhouse gas emissions, while simultaneously consuming valuable resources. Fortunately, much of the recoverable heavy crude oil is located in areas with high solar irradiance. Thus STT generated steam becomes a viable option for EOR, more so in places where the cost of natural gas is high. A solar steam demonstration project based on solar tower technology has been set up at Chevron's Coalinga field in California. With the help of one ~100 m tall tower and 3822 heliostats, 60% quality steam is generated at 260 °C and 48.3 bar [7].

4 Thermoelectric Co-generation

Researchers at MIT have adapted the conventional parabolic trough collector (PTC) to come up with a system that simultaneously produces "generatorless" electricity and heat [8]. In a conventional PTC, the inner steel tube, which contains the thermic fluid, is coated with a solar selective coating with high absorptivity and low emissivity. The metal tube, in turn, is surrounded by a glass tube; the annular space between them contains vacuum to minimize convection losses. The adapted, MIT system is depicted in Fig. 4. The collector, inclined at an angle θ, consists of three concentric tubes, with the innermost tube containing the thermic fluid.

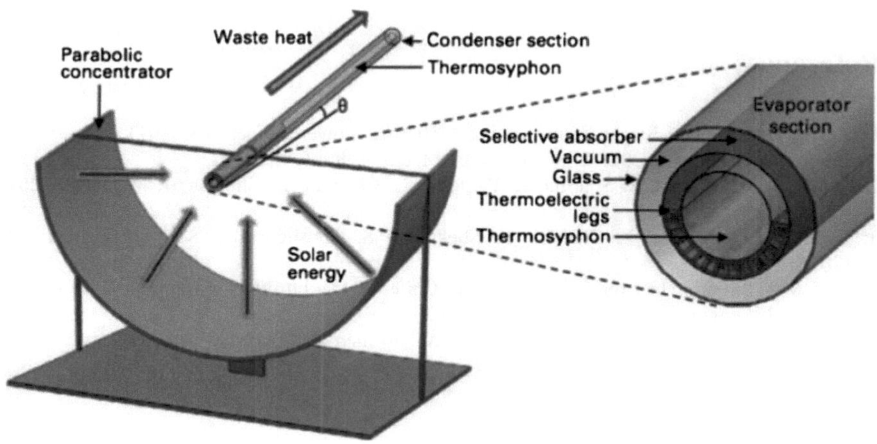

Fig. 4 A modified parabolic trough system for thermoelectric co-generation

The middle tube has a solar selective coating; its surface acquires the highest temperature on exposure to concentrated solar irradiance. A vacuum exists between the two outer tubes. Themoelectric (TE) devices fill the space between the two inner tubes with one leg in contact with the "hotter" central tube and the other in contact with the relatively "colder" innermost tube. The higher end of the collector extends beyond the width of the parabolic mirror; this end is relatively cooler as the solar radiation from the parabolic mirror does not fall on it. When concentrated solar radiation from the parabolic mirrors falls on the collector, the TE device generates electricity. Simultaneously, the thermic fluid is vapourized and the vapour flows towards the cooler, upper end, where it releases heat, condenses and flows down to the lower end by gravity. This process is continuously repeated, creating a thermosyphon. This system may be of great use for distributed co-generation.

5 Conclusions

STT is very versatile, which can simultaneously produce electricity, heating and cooling. In fact, the proportion of three outputs can, in principle, be changed depending on the requirements, like a three-way valve. There are other potential innovative uses of STT. Take the example of cold storages in India, which are largely located in semi-urban and rural areas where electricity supply is very erratic. Consequently, a significant proportion of India's fruits and vegetables perish. A composite, self-compensating vapour absorption-compression system could be a possible solution. Despite the current high costs of STT, because of its versatility, I expect STT to form a significant proportion of the renewable portfolio in the near future. And if you add to it, doses of human ingenuity and innovation, sky would be the limit for STT.

References

1. K. De Decker, The Bright Future of Solar Thermal Powered Factories (2011). http://www.lowtechmagazine.com/2011/07/solar-powered-factories.html
2. D. Patidar, S. Tiwari, P. Sharma, R. Pardeshi, L. Chandra, R. Shekhar, J. Metals **67**, 2696 (2015)
3. P. Sharma, R. Sarma, L. Chandra, R. Shekhar, P.S. Ghoshdastidar, Sol. Energy **111**, 135 (2015)
4. 27.5 MW Provide Heat for Copper Mine in Chile (2015). http://www.solarthermalworld.org/content/275-mw-provide-heat-copper-mine-chile
5. A. Z'Graggen, P. Haueter, D. Trommer, M. Romero, J.C. de Jesus, A. Steinfeld, Int. J. Hydrogen Energy **31**, 797 (2006)

6. M. Dolan, Solar Hybrid Fuels (Stream 1): Syngas from Natural Gas at Lower Temperatures Compatible with Conventional Solar Thermal Storage, While Maintaining High Efficiency (2016). http://arena.gov.au/files/2016/05/Solar-Hybrid-Fuels-Stream-1-Summary-report-project-results-and-lessons-learnt.pdf
7. A.R. Kovscek, J. Petrol. Sci. Eng. **98–99**, 130 (2012)
8. N.W. Stauffer, Harnessing Solar Energy: Novel Approach Harnesses Both Electricity and Heat (2012). http://news.mit.edu/2012/harnessing-solar-energy

Comparison of Thermal Characteristics of Sensible and Latent Heat Storage Materials Encapsulated in Different Capsule Configurations

Muthukumar Palanisamy and Hakeem Niyas

Abstract This paper presents the thermal modeling and performance comparison of sensible and latent heat based thermal energy storage (TES) systems using concrete and phase change materials (PCMs) encapsulated in containers of different geometrical configurations. The sensible heat storage (SHS) and latent heat storage (LHS) module considered here is a capsule containing concrete or sodium nitrate which exchanges heat with the source material. SHS capsule is modeled using the energy conservation equation. Effective heat capacity method is employed to account the latent heat of the PCM. Boussinesq approximation and Darcy law's source term are added in the momentum equation to incorporate the natural convection of molten PCM and nullify the velocities of solid PCM. The equations of the 2D axisymmetric model are solved using COMSOL Multiphysics. Charging time of capsules in four different configurations viz., spherical, cylindrical ($H = D$, $H = 4D$) and novel cylindrical configurations are compared. The thermal characteristics are compared using isothermal contour plots and temperature–time curves.

Keywords Encapsulation · Performance prediction · Solar thermal · Energy storage

1 Introduction

Several countries ramped up their investments on renewable energy based electricity generation to combat the harmful environmental outcomes from power plants running on fossil fuels. Indeed the availability of fossil fuels is also a big question after few decades. Concentrated solar power (CSP) is one of the promising large-scale power generation technologies among the renewables which is being widely commercialized now. The major problem that CSP plants face is the

M. Palanisamy (✉) · H. Niyas
Department of Mechanical Engineering, Indian Institute of Technology Guwahati, Guwahati, Assam, India
e-mail: pmkumar@iitg.ernet.in; pmkumariitg@gmail.com

L. Chandra and A. Dixit (eds.), *Concentrated Solar Thermal Energy Technologies*, Springer Proceedings in Energy, https://doi.org/10.1007/978-981-10-4576-9_2

intermittent solar radiation which limits the capacity and reliability of the plant. To alleviate this, integration of thermal energy storage (TES) systems to the CSP plants is essential [1]. Till date, two TES technologies were implemented commercially, viz., molten salt systems for parabolic trough or solar tower CSP plants and steam accumulators for direct steam generation CSP plants [2]. The major disadvantage of steam accumulators is the decrease in pressure during the discharging of steam. Incorporating a flash evaporator or an encapsulated heat storage capsule is preferred to avoid or reduce the pressure drop. Due to entropy generation by mixing in the flash evaporator system, encapsulated capsule is beneficial in steam accumulators [3]. Thermal energy can be stored in the form of sensible heat, latent heat and thermochemical heat. Sensible and latent heat storage devices are viable options for usage in steam accumulators in the form of encapsulated capsules. SHS devices store the sensible heat of the material during the rise in temperature. LHS devices store the latent heat of phase change material (PCM) at near constant melting temperature. But thermal resistance during charging/discharging of certain SHS material like concrete and PCM is high because of its low thermal conductivity. Several techniques were analyzed by researchers to increase the performance of the storage systems [4–6]. A detailed review of performance improvement techniques for was reported in the literature [7]. Encapsulating storage materials inside capsule increases the specific surface area. Also, direct contact of heat transfer fluid with the capsule increases the heat transfer coefficient [8]. Spherical and cylindrical capsules filled with PCM were tested for storage characteristics in a constant temperature bath at lab scale and a steam accumulator at industrial scale [9–13]. It is understood from the literature survey that no work was reported comparing the performance characteristics of different configurations of encapsulated capsules. In the present work, a numerical model is developed for evaluating the performances of different configurations of encapsulated capsules. Concrete and sodium nitrate packed in four different configurations, viz., spherical, cylindrical ($H = D$, $H = 4D$), and novel cylindrical models are compared.

2 Model Description

Figure 1 shows a view of the 2D axisymmetric encapsulated capsule containing concrete or sodium nitrate. The capsules are designed for a heat storage capacity of 320 kJ each. Melting point, latent heat, and thermal conductivity are the major parameters which are having a great influence in the design of any LHS systems. Similarly, specific heat and thermal conductivity of the storage material decides the charging time of SHS systems. The dimensions of the capsules are given in Table 1. Encapsulating material should withstand the high pressures (60–80 bar) that exists inside the pressure vessel. SS304 is selected as the encapsulating material for the present analysis. The thermophysical properties of concrete, sodium nitrate and SS304 are given in Table 2. Three physical processes are to be simulated to study the thermal storage behavior of encapsulated LHS capsules, i.e. conduction,

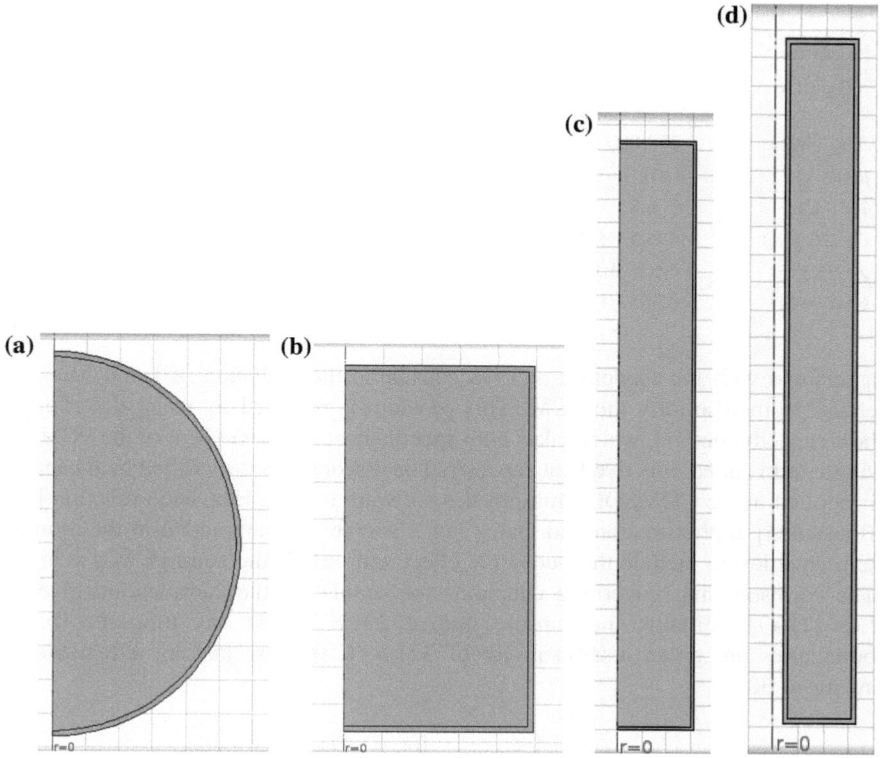

Fig. 1 2D axisymmetric view of capsules **a** spherical **b** cylindrical, $H = D$ **c** cylindrical, $H = 4D$ and **d** novel cylindrical

Table 1 Dimensions of capsules

Configuration	Concrete		Sodium nitrate	
	D (mm)	H (mm)	D (mm)	H (mm)
Spherical	223	–	110	–
Cylindrical ($H = D$)	195	195	96	96
Cylindrical ($H = 4D$)	122.5	490	60.5	242
Novel cylindrical	123	490	61	242

convection and phase change. Conduction heat transfer is the only process involved in the charging simulation of SHS capsules. A 2D axisymmetric model is developed in view of the symmetry of flow and heat transfer around the vertical axis. Molten PCM movement within the capsules due to natural convection heat transfer is assumed to be laminar, Newtonian and incompressible. The major problem

Table 2 Thermo-physical properties of sodium nitrate, concrete and SS304

Properties	Sodium nitrate [15–18]	Concrete [19]	SS304 [19]
ρ (kg/m^3)	2130 (solid) 1908 (liquid)	2300	8000
ΔH_F (J/kg)	178,000	–	–
μ (Pa s)	0.0119–1.53 × 10^{-5} T	–	–
T_M (°C)	306.8	–	–
C_P (J/kg K)	444.53 + 2.18 T	800	500
α (1/K)	6.6 × 10^{-4}	1 × 10^{-5}	1.78 × 10^{-5}
k (W/m K)	0.3057 + 4.47 × 10^{-4} T	1.2	16.2

associated with the modeling of LHS capsule is the inclusion of latent heat necessary to melt/solidify the PCM. This problem is resolved by using the effective heat capacity method, which takes both specific heat and latent heat of the PCM in a single term called effective heat capacity. The discontinuous modified heat capacity is applied in the COMSOL Multiphysics software using a Heaviside function [14]. Boussinesq approximation and Darcy law's source term are added in the momentum equation to include the buoyancy effect and nullify the solid PCM's velocity. The corresponding governing equations for simulating the capsules are given in Eqs. (1)–(7). Initially, the capsules are at 291.8 °C. At any time ($t > 0$), the boundaries are given a temperature of 321.8 °C thereby making a temperature swing of 30 °C

$$C_P = \begin{cases} C_{PS} & \text{for } T < T_S \\ C_{P,\mathrm{EFF}} & \text{for } T_S \le T \le T_L \\ C_{PL} & \text{for } T > T_L \end{cases} \tag{1}$$

$$C_{P,\mathrm{EFF}} = \frac{C_{PS} + C_{PL}}{2} + \frac{\Delta H_F}{2\Delta T_M} \tag{2}$$

$$\nabla.\vec{V} = 0 \tag{3}$$

$$\frac{\partial \vec{V}}{\partial t} + (\vec{V}.\nabla)\vec{V} = \frac{1}{\rho}\left(-\nabla P + \mu \nabla^2 \vec{V} + F + \vec{S}\right) \tag{4}$$

$$\rho C_p \frac{\mathrm{D}T}{\mathrm{D}t} = k \nabla^2 T \tag{5}$$

$$F = \rho \vec{g} \beta (T - T_M) \tag{6}$$

$$\theta = \frac{T - T_S}{T_L - T_S} = \frac{T - T_M + \Delta T_M}{2\Delta T_M} = \begin{cases} 0 & \text{for } T < T_S \\ 0 - 1 & \text{for } T_S \le T \le T_L \\ 1 & \text{for } T > T_L \end{cases} \tag{7}$$

3 Results and Discussions

3.1 Validation

In order to validate the numerical model, the results obtained for the melt fraction of storage bed were compared with the melt fraction reported in the literature [10]. The physical model chosen for the numerical validation, thermophysical properties of the PCM and initial and boundary conditions of the model are taken from literature [10]. It is evident from Fig. 2 that the current numerical results showed good agreement with the results reported in the literature [10].

3.2 Grid Independency Test

Free triangular mesh has been adapted in the numerical model. In order to test the dependency of numerical results on the mesh element size, a simulation is run with the cylindrical capsule of $H = D$. The LHS capsule is initially at 291.8 °C. At any time $t > 0$, the boundary of the capsule is at 321.8 °C. The average temperature of the capsule is compared for different element sizes and it is observed from Fig. 3 that 10,377 elements can be taken for the numerical model. Similarly, grid independency test carried has been carried out for other configurations also. Time step used in the analyses is 0.01 s throughout all the models.

3.3 Temperature Distribution

Figure 4a, b shows the comparison of the average temperature variation of the LHS/SHS capsules of different configurations. During charging, PCM/concrete kept in the capsules is initially in the solid state at 291.8 °C. When a high temperature of

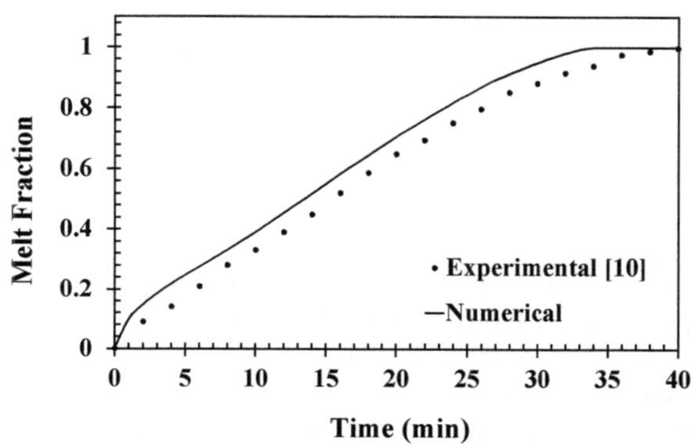

Fig. 2 Validation of the numerical model

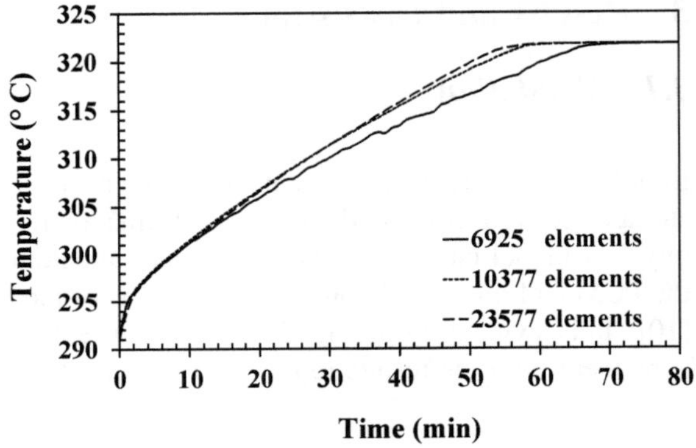

Fig. 3 Grid independency test

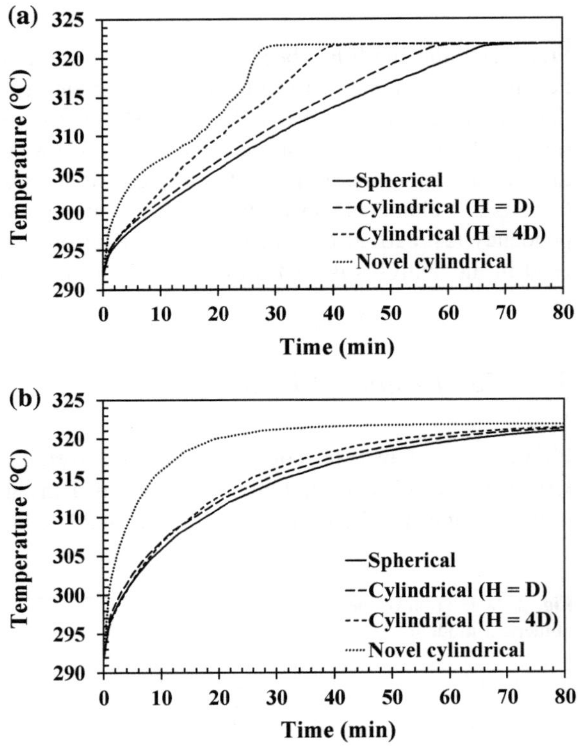

Fig. 4 Temperature variation **a** LHS and **b** SHS capsules

321.8 °C is applied on the boundary of the capsule, heat is transferred from periphery of the capsules and thereby stored in the form of sensible and latent heat. It is inferred from Fig. 4 that the increase in average temperature is faster in cylindrical capsules than the spherical capsule. Also, it can be noted that the cylindrical capsule with a higher aspect ratio reaches the boundary condition in

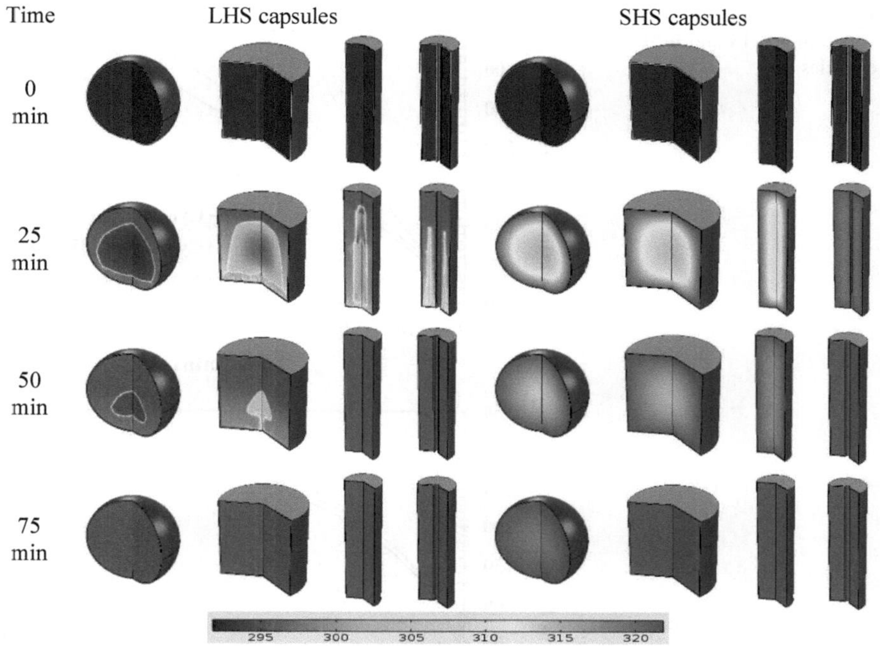

Fig. 5 Temperature variation of LHS and SHS capsules

lesser time. Also, the increase in temperature of novel cylindrical configuration is much faster than all of the configurations. This is due to two reasons, viz., increased heat transfer area and reduced thickness of the capsule. Screenshots of temperature variation of LHS capsules during charging are given in Fig. 5. The effect of natural convection is seen from the asymmetric temperature variation in all the LHS capsules. Similarly, the conduction in the concrete capsules is seen from the symmetric variation of temperature.

3.4 Total Energy Storage Rate

Figure 6a, b show the total energy storage rate of all the LHS/ SHS capsules. For making a comparison between the LHS and SHS capsules of different configurations, the time taken for 95% of total energy stored, i.e., 304 kJ can be taken as the charging time of the capsules. It took about 58, 48, 36, and 26 min to store about 320 kJ in the LHS capsules. Similarly, it took about 69, 62, 57, and 22 min to store about 320 kJ in the SHS capsules.

In the SHS capsules, minor difference in the charging time exists between the spherical, cylindrical ($H = D$), and cylindrical ($H = 4D$) configurations. But there

Fig. 6 Total energy storage variation **a** LHS and **b** SHS capsules

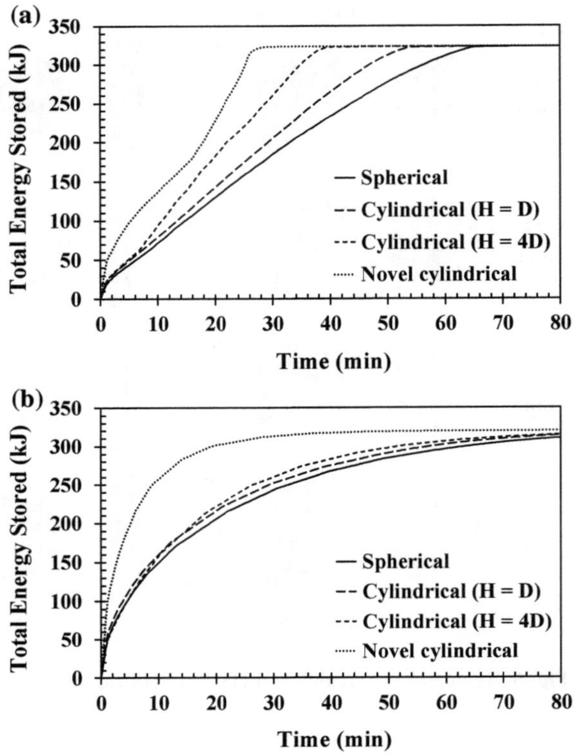

exists a huge decrease in the charging time using the novel cylindrical configuration. In the LHS capsules, considerable difference in the charging time exists between the spherical, cylindrical ($H = D$), cylindrical ($H = 4D$), and novel cylindrical configurations. This is due to the presence of natural convection in the capsules during melting. The proposed novel cylindrical configuration yields 61 and 28% lesser charging times for SHS and LHS capsules when compared with the cylindrical capsule ($H = 4D$).

4 Conclusions

A thermal model was developed to compare the performance of LHS/SHS capsules of different geometrical configurations. Numerical results shown that for the same mass of storage material, cylindrical configuration yields lesser charging time than the spherical configuration. This is due to the fact that the distance between the center and periphery of the capsule is more in spherical capsule than that of cylindrical capsule. In the cylindrical models, configuration with a higher aspect ratio takes lesser time for charging. The time taken for charging of LHS capsule in spherical, cylindrical ($H = D$, $H = 4D$) and novel cylindrical configurations are 58, 48, 36, and 26 min, respectively. Similarly, the time taken for charging of SHS capsule in spherical, cylindrical ($H = D$, $H = 4D$), and novel cylindrical configurations are 69, 62, 57 and 22 min, respectively. The proposed novel cylindrical configuration saves 61 and 28% charging times for SHS and LHS capsules when compared with the cylindrical capsule ($H = 4D$). The results of the current study will be useful in the design and optimization of storage modules which are having wide usage in solar thermal power plants, clean transportation systems, thermal conditioning for buildings, etc.

Acknowledgements The authors sincerely thank the Department of Science and Technology (DST), Government of India, for the financial support (Project No: DST/TM/SERI/2K10/53(G)).

References

1. S.H. Madaeni, R. Sioshansi, P. Denholm, Capacity value of concentrating solar power plants. National Renewable Energy Laboratory Technical Report, NREL/TP-6A20-51253 (2011)
2. C. Prieto, A. Jove, F. Ruiz et al., in *Commercial Thermal Storage. Molten Salts vs. Steam Accumulators*. Proceedings, SolarPACES. http://www.solarpaces2012.org (2012)
3. W.D. Steinmann, M. Eck, Buffer storage for direct steam generation. Sol. Energy **80**, 1277–1282 (2006)
4. I.T. Barney, S. Ganguli, A.K. Roy et al., Improved thermal response in encapsulated phase change materials by nanotube attachment on encapsulating solid. J. Nanotechnol. Eng. Med. **3**, 6–0310051 (2012)
5. A. Sari, A. Karaipekli, Thermal conductivity and latent heat thermal energy storage characteristics of paraffin/expanded graphite composite as phase change material. Appl. Therm. Eng. **27**, 1271–1277 (2007)
6. L.C. Chow, J.K. Zhong, J.E. Beam, Thermal conductivity enhancement for phase change storage media. Int. Commun. Heat Mass Transfer **23**(1), 91–100 (1996)
7. L. Liu, W. Saman, F. Bruno, Review on storage materials and thermal performance enhancement techniques for high temperature phase change thermal storage systems. Renew. Sustain. Energy Rev **16**, 2118–2132 (2012)
8. A. Mathur, R. Kasetty, J. Oxley et al., Using encapsulated phase change salts for concentrated solar power plant. Energy Procedia **49**, 908–915 (2014)
9. C.W. Chan, F.L. Tan, Solidification inside a sphere—an experimental study. Int. Commun. Heat Mass Transfer **33**, 335–341 (2006)
10. E. Assis, L. Katsman, G. Ziskind et al., Numerical and experimental study of melting in a spherical shell. Int. J. Heat Mass Transfer **50**, 1790–1804 (2007)
11. A.R. Archibold, M.M. Rahman, D.Y. Goswami et al., Analysis of heat transfer and fluid flow during melting inside a spherical container for thermal energy storage. Appl. Therm. Eng. **64**, 396–407 (2014)
12. N. Himeno, K. Hijikata, A. Sekikawa, Latent heat thermal energy storage of a binary mixture-flow and heat transfer characteristics in a horizontal cylinder. Int. J. Heat Mass Transfer **31**(2), 359–366 (1988)
13. W.D. Steinmann, R. Tamme, Latent heat storage for solar steam systems. J. Sol. Energy Eng. **130**, 5–0110041 (2008)
14. COMSOL Inc., http://www.comsol.com/

15. T. Bauer, D. Laing, R. Tamme, Characterization of sodium nitrate as phase change material. Int. J. Thermphys. **33**, 91–104 (2012)
16. G.J. Janz, C.B. Allen, N.P. Bansal et al., Physical Properties Data Compilations Relevant to Energy Storage. II. Molten Salts: Data on Single and Multi-component Salt Systems. NSRDS-NBS-61. http://www.nist.gov/data/nsrds/NSRDS-NBS61-II.pdf (1979)
17. C.W. Lan, S. Kou, Effects of rotation on heat transfer, fluid flow and interfaces in normal gravity floating-zone crystal growth. J. Cryst. Growth **114**, 517–535 (1991)
18. L.R. White, H.T. Davis, Thermal conductivity of molten alkali nitrates. J. Chem. Phys. **47**, 5433–5439 (1967)
19. AZo Network UK Ltd., http://www.azom.com/

Part II
Solar Field

Development of Dual-Axis Tracking Control System for Scheffler Dish Concentrator

A. Gokul Raj, Arun Bollavarapu and Mukul Dubey

Abstract Scheffler dish solar concentrating systems are used in cooking or process heating applications requiring temperature up to 300 °C. These concentrators are generally equipped with tracking facility to track the sun's path throughout the day to obtain direct beam radiation. The tracking system helps in achieving the effective operation of the system by concentrating the direct beam radiation at a focal point of the receiver. Currently, in Scheffler dish solar concentrating systems, single-axis tracking, i.e., East–West (E–W) is achieved using timer device while tilting axis, i.e., North–South (N–S) is achieved by adjusting the bars manually. The timer device is either pendulum regulated mechanical system or expensive and intricate electronic timer circuit. A simple and economically viable electronic control for Scheffler dish concentrating systems is highly desired. This paper discusses a simplified and cost-effective circuit design to achieve dual-axis tracking (both E–W and N–S) for easy and effective operation of Scheffler dish concentrator.

1 Introduction

Dual-axis tracking system is essential for concentrating solar power (CSP) technologies such as Scheffler dish to harness maximum solar energy by tracking daily and seasonal variations of solar positions. The commercially available electro-optical trackers are accurate but expensive due to intricate design and sophisticated fabrication needs.

Barsoum [1] used a chain drive with 4:1 gear reduction ratio between motor socket and frame socket, Khan et al. [2] developed a tracking system with 150° total angle of rotation (east to west) with 3.75° rotation of stepper motor for every 15 min and Saravanan et al. [3] integrated a three-phase 0.5 HP induction motor to their solar tracking system with gear reduction ratio of 25:1, where we used a gear reduction

A. Gokul Raj (✉) · A. Bollavarapu · M. Dubey
Sardar Patel Renewable Energy Research Institute, Vallabh Vidhyanagar 388120, Gujarat, India
e-mail: gokulraj1986@gmail.com

L. Chandra and A. Dixit (eds.), *Concentrated Solar Thermal Energy Technologies*, Springer Proceedings in Energy, https://doi.org/10.1007/978-981-10-4576-9_3

drive of 30:1 ratio to achieve smooth and controlled tracking. Abdul-lateef et al. [4] developed a single-axis tracker prototype with PIC16F84A flash microcontroller and 9 V permanent magnet geared DC motor of lower capacity for lighter loads, Kansal [5] developed a single-axis sun tracking system using PIC microcontroller and stepper motors and achieved 8% tracking efficiency at field conditions.

In our work we employ inexpensive DC wiper motor that has the capability to withstand higher loads. Further, in contrast to Patil et al. [6] we are using a low-cost timer circuit.

2 Experimental Setup

The Scheffler dish concentrator system consists of a reflector and receiver. The reflector forms a lateral section of a paraboloid and the receiver is placed at its focal point. In our setup, for reflector we used polished aluminium sheets which were firmly inserted between aluminium sections along with rigid frame and stand made of mild steel. Boiler grade mild steel tank with black-coated face was used as receiver. The face of the receiver absorbs the concentrated rays from the reflector and transfer the heat energy to the water in the tank to produce steam. The saturated steam at desired pressure is obtained using a resettable pressure release valve. Two linear actuators were fitted with the receiver to achieve seasonal tracking. The schematic of the Scheffler dish concentrator is shown in Fig. 1.

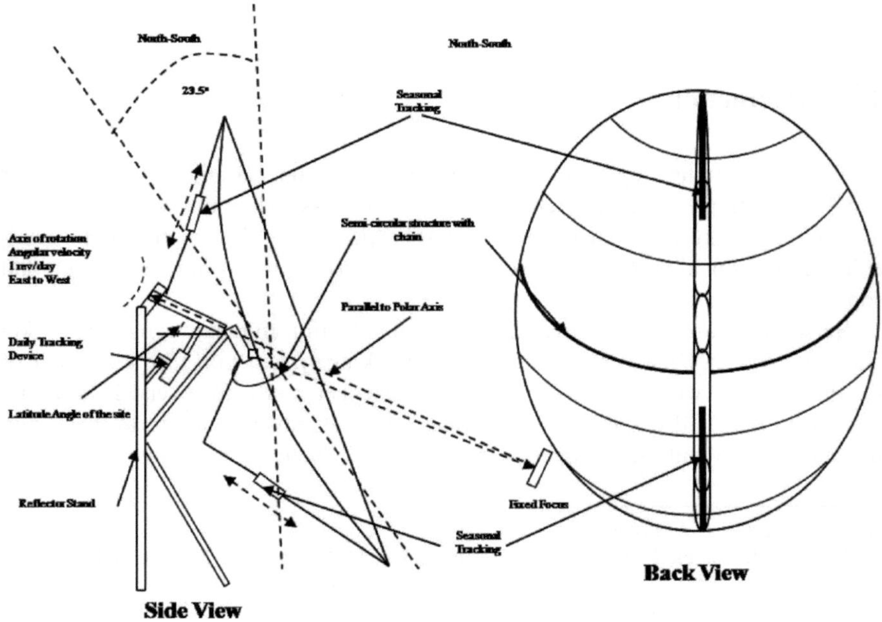

Fig. 1 Schematic views of the Scheffler dish concentrator

3 Design and Development of Tracking System

3.1 East–West Tracking Control

The semicircular structure of the Scheffler dish is used for achieving smooth and controlled tracking. A chain slides over the semicircular structure. A compound gear reduction drive is connected between the DC motor and the chain through sprocket. The operation of DC motor is controlled by the timer control circuit.

The control circuit for E–W tracking comprises step-down transformer (230 V AC–12 V AC supply), rectifier (12 V AC–12 V DC), DC voltage regulator (for constant 5 V DC), PIC 16F876 micro controller, LED display, LED indicators, relays, and control switches. An AC-to-DC adaptor of 12 V, 5 A is used to supply power to the low-cost motor.

The schematic block diagram of the E–W solar tracking system is shown in Fig. 2. The design features of the tracking system are as follows.

A PIC microcontroller and other electronic components used in the control unit were procured locally. A compound gear reduction drive having gear reduction ratio of 30:1 was procured from a local supplier. The program was written in PICBASIC. The program code is compiled and assembled using PB-PLUS compiler. The usage of battery was eliminated by using an AC-to-DC adaptor. A simple circuit with fewer electronic components and single-layer PCB design has been developed using Proteus Design Suite. The circuit diagram is shown in Fig. 3.

The DC motor is connected to the output pin 7 of the microcontroller. Push buttons S2 and S3 are used for ON time increments and ON time decrements. Push buttons S4 and S5 are used for OFF time increments and OFF time decrements. Push buttons S5 and S1 are used to start ON–OFF cycle and to reset ON–Off cycle. The ON–Off cycle should be set accordingly to achieve 15° angle of rotation per hour. The cycle continues throughout the day to achieve E–W tracking.

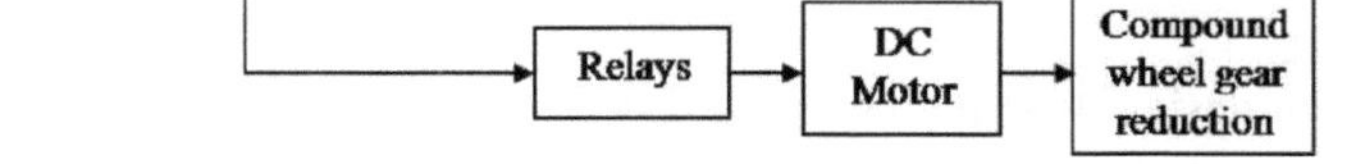

Fig. 2 Schematic block diagram E–W solar tracking system

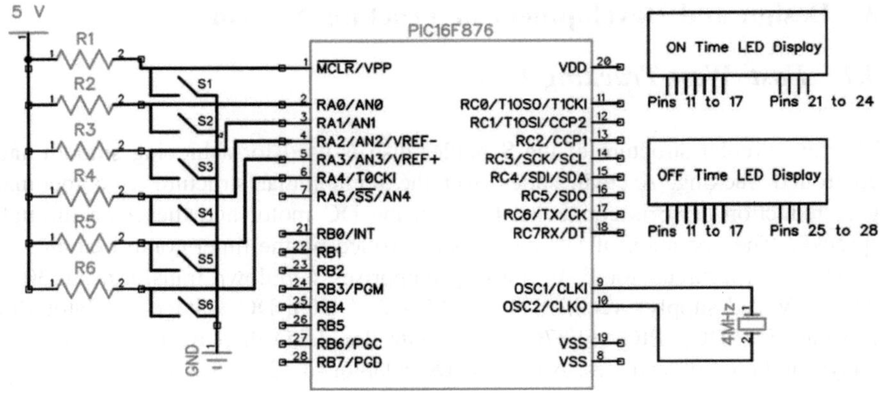

Fig. 3 Circuit diagram for east–west tracking control of Scheffler dish

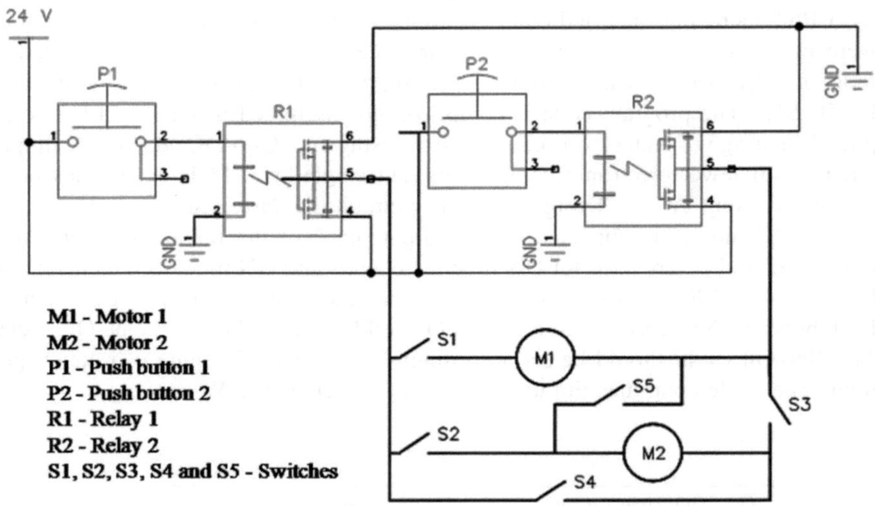

Fig. 4 Circuit diagram for north–south control of Scheffler dish

3.2 North–South Tracking Control

The circuit design for north–south control of the Scheffler dish was developed, as shown in Fig. 4.

The circuit design was simulated and tested in Proteus for the control operation. P1 and P2 are used for controlling the relays R1 and R2, respectively. Switches 1–5 are used for achieving different control configurations of the motors M1 and M2 when P1 and P2 are activated. After fabrication of the circuit in PCB, the circuit will be connected to two linear actuators of 24 V and 4 A rated capacity to achieve the seasonal adjustment (N–S tracking).

4 Simulation Results and Discussion

The simulation of the designed E–W and N–S circuits were carried out in Proteus simulation tool. The working of the E–W control circuit was verified by connecting the 7th pin (RA5) of the microcontroller to a NO (normally open) relay RL having excitation coil voltage of 5 V, 10 A maximum allowable current at contact terminal and connecting the 12 V DC motor through the relay contact (Fig. 5).

The program for the cyclic timing operation was fed into the microcontroller. The ON and OFF cycle operation of the relay was tested with a sample timing cycle of 5 s ON and 30 s OFF. It was found that the relay operated in accordance with the ON–OFF cycle. In Fig. 6, the blue line peeks indicate the ON cycle voltage at relay

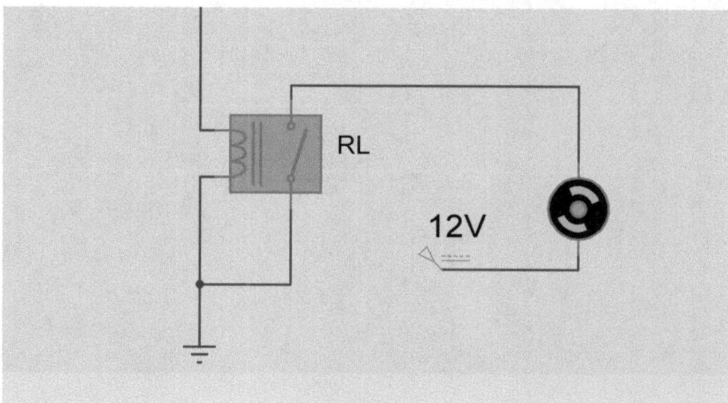

Fig. 5 Connection showing the 12 V DC motor with the relay contact

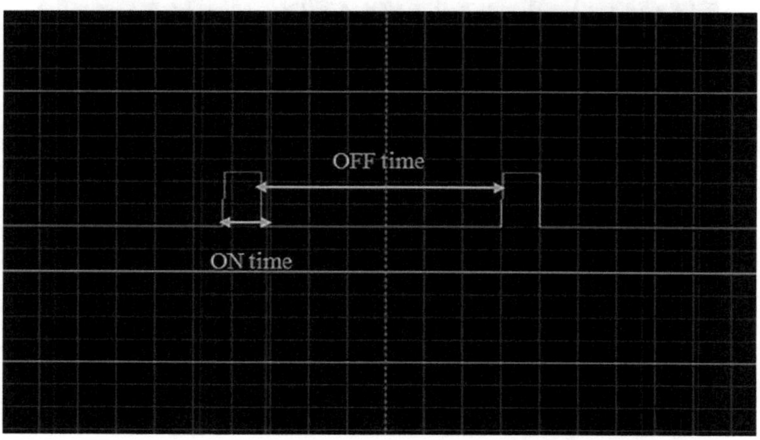

Fig. 6 Simulation result of relay ON OFF cycle for E–W tracking

contact. Thus, the E–W tracking can be achieved by connecting the supply to the 12 V DC motor through the relay RL. The ON–OFF cycle to achieve 15° E–W angular rotation per hour of the Scheffler dish concentrator is 1 and 74 s respectively.

The working of the N–S control was also verified for different switching configurations. In Fig. 7, the yellow and green line peeks indicate the forward and reverse currents of the motors M1 and M2 when push button P1 is pressed while

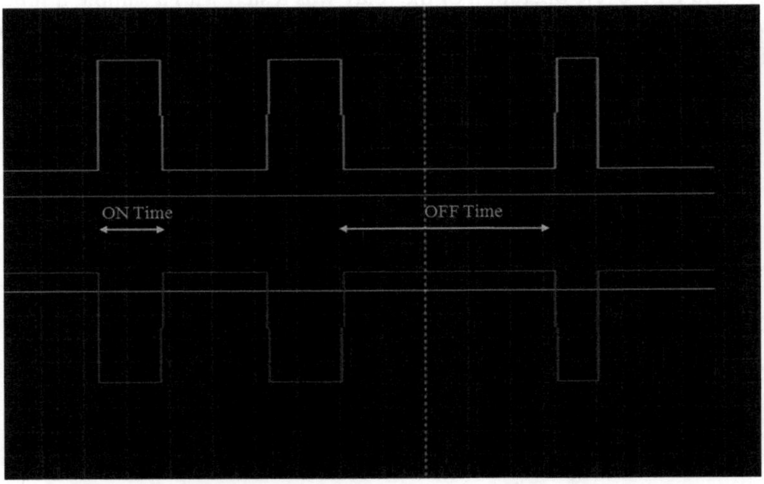

Fig. 7 Simulation result showing the forward and reverse operation of the motors M1 and M2 for N–S tracking when push button P1 is pressed

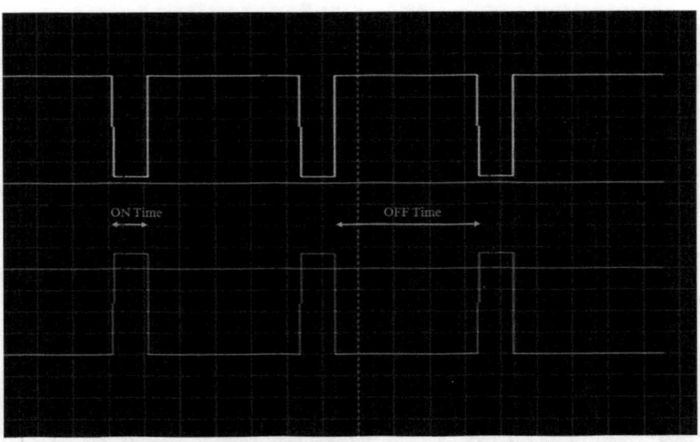

Fig. 8 Simulation result showing the reverse and forward operation of the motors M1 and M2 for N–S tracking when push button P2 is pressed

Fig. 8, the yellow and green line peeks indicate the reverse and forward currents of the motors M1 and M2 when push button P2 is pressed, indicating successful operation of both forward–reverse and reverse–forward rotations.

5 Conclusion

Simulation results indicated that a timer based electronic control circuit can be used to achieve dual-axis tracking (both E–W and N–S) of Scheffler dish concentrator through PIC microcontroller and relays. The circuit design is simple and uses inexpensive components making it a cost-effective choice both in terms of fabrication and maintenance. The future work will include fabrication of circuit on PCB and its comprehensive evaluation.

Acknowledgements The work is funded by Department of Science and Technology, Government of India. The authors sincerely acknowledge Dr. M. Shyam (Director, SPRERI) for his support and guidance. The authors also acknowledge Dr. V. Siva Reddy (Ex-Head of the Department) for initiating the work and providing valuable guidance.

References

1. N. Barsoum, Implementation of dual-axis solar tracking pilot project. Trans. Energy Biotechnol. Plann. Environ. ET-E33/GJTO, 2229–8711 (2011)
2. M.T.A. Khan, S.M. Shahrear Tanzil, R. Rahman, S.M. Shafiul Alam, in *Design and Construction of an Automatic Solar Tracking System*. 6th International Conference on Electrical and Computer Engineering ICECE (2010)
3. C. Saravanan, M.A. Panneerselvam, I. William Christopher, A Novel low cost automatic solar tracking system. Int. J. Comput. Appl. **31**, 0975–8887 (2011)
4. K.I. Abdul-lateef, A low cost single-axis sun tracker system using PIC microcontroller. Diyala J. Eng. Sci. **5**, 65–78 (2012)
5. R. Kansal, PIC Based Automatic Solar Radiation Tracker (M. Tech thesis). Thapar University, Patiala, Punjab (2008)
6. R.J. Patil, G.K. Awari, M.P. Singh, Comparison of performance analysis of Scheffler reflector and model formulation. Int. J. Sci. Technol. 4(10) (2011)

Comments on Quality Control of Solar Radiation Data Measured at a Ground Station in Hot and Dry Zone

D. Kumar and B. Ravindra

Abstract Ministry of New and Renewable Energy, Govt. of India has set a target of 100 giga watts (GW) power from solar power till 2022. Hence, the importance of solar radiation resource assessment. Ground-based solar radiation measurement and its quality plays an important role in the study of estimation of solar radiation and forecasting of solar power produced. In this paper hourly averaged solar irradiance data on a horizontal surface, measured at Indian Meteorological Department weather station for the hot and dry region, Jodhpur (India) is analyzed for the month of August, 2015. Data quality check is done in two stages where, at the first stage, solar irradiance is taken in original form, and then in the second stage BSRN standard guidelines, coherence and correlation tests are applied. Data quality check of ground-based solar radiation data for Jodhpur during rainy season (month having highest error/disturbances) highlights the need to include other weather parameters such as rainfall data. The results are compared with other available data, and some suggestions for data quality improvement are discussed.

Keywords IMD · Irradiance · NREL SERI-QC · BSRN

Symbols

ε Earth Eccentricity Factor
I_E Extraterrestrial Radiation
W Watts
Z Zenith Angle

D. Kumar (✉)
Centre for Energy, Indian Institute of Technology Jodhpur, Jodhpur 342001, Rajasthan, India
e-mail: pg201281002@iitj.ac.in; dharm.career@gmail.com

B. Ravindra
Department of Mechanical Engineering, Indian Institute of Technology Jodhpur, Jodhpur 342001, Rajasthan, India

L. Chandra and A. Dixit (eds.), *Concentrated Solar Thermal Energy Technologies*, Springer Proceedings in Energy, https://doi.org/10.1007/978-981-10-4576-9_4

Abbreviations

AMS	Advance Measurement Station
BSRN	Baseline Surface Radiation Network
C-WET	Centre for Wind Energy Technology
CIEMAT	Centro de Investigaciones Energéticas, Medioambientales y Tecnológicas
DHI	Diffuse Horizontal Irradiance (W/m^2)
DLR	Deutsches Zentrum Für Luft- und Raumfahrt
DNI	Direct Normal Irradiance (W/m^2)
ETR	Extraterrestrial Radiation
GHI	Global Horizontal Irradiance (W/m^2)
IMD	Indian Meteorological Department
ISO	International Organization for Standards
JNNSM	Jawaharlal Nehru National Solar Mission
K	Diffuse Horizontal Transmittance (DHI/GHI)
K_n	Direct Beam Transmittance (DNI/I_E)
K_t	Clearness Index (GHI/(I_E x cos(z)))
MESOR	Management and Exploitation of Solar Resource Knowledge
MNRE	Ministry of New and Renewable Energy
NIP	Normal Incidence Pyrheliometer
NREL	National Renewable Energy Laboratory
PSP	Precision Spectral Pyranometer
RMIB	Royal Meteorological Institute of Belgium
SERI-QC	Solar Energy Research Institute Quality Control software package
SRRA	Solar Radiation Resource Analysis
WMO	World Meteorological Organization

1 Introduction

Due to rapid consumption of fossils fuels and huge demand for power, MNRE, Govt. of India, has introduced a solar power generation plan named Jawaharlal Nehru National Solar Mission (JNNSM) [1]. Currently, the ministry has updated its solar target to 100 GW till 2022 [2]. To achieve this target, solar radiation resource assessment is needed. The solar radiation data can be obtained from several sources. The first one is the IMD-Pune (National Data Centre) [3], which is around for the last 100 years, from where we can take satellite as well as ground data. Next is C-WET Chennai which has installed 111 ground-based SRRA stations and 4 AMS which provide solar radiation data with very fine frequency (1 min). To achieve the given solar power generation target mentioned above, the most important knowledge required is good quality solar radiation data, with no missing values and free from all errors.

In this paper, the solar radiation data quality is demonstrated for one Indian location. It is known that climate classification may also play a role in solar resource assessment. Here we have taken classification provided in MNRE handbook: Energy Conscious Buildings [4]. This is based on the work of Bansal et al. [5] who have proposed five climatic zones [6].

For any location-specific analysis, the first thing required is the long-term time-series solar radiation data of that location. The available data can be of three types: satellite data, time-series data, and ground-based data. The satellite data is basically used for preliminary site analysis, but for any detailed assessment, ground-based data is required for at least 1 year [7]. In this research article, we will use only ground-based solar radiation data.

The accurate measurement of ground-based solar radiation data requires great care. The instruments required for solar measurement: Pyranometers for the DHI and GHI, and Pyrheliometer for DNI, require careful operation and maintenance and only best class instruments which follow (ISO-9060) [8] are selected for actual research grade databases. On-line data checking is required for getting best class data. Such tests routinely used for checking the data quality will be discussed later in this article.

The reasons behind errors occurring in solar radiation measurement are equipment error, uncertainty, and operational errors [9]. On equipment side, the errors usually are cosine and azimuth response, temperature response etc. out of which maximum error is caused by cosine error. Usually all these errors are controlled and checked by manufacturer and by the calibration center. A routine check-up is done for instrument's sensitivity and its quality standard, which certifies them for further use on ground. Reading, cleaning, alignment or other operational errors such as natural errors are also not uncommon.

To study the effects of natural errors Gender and Quashing [10] have worked on the response of dust/sand layers on measuring instruments and made some suggestions for correction of data. They suggested use of Si-based sensors, if the site is in a remote place and regular operation and maintenance is not possible. It may be noted that poor quality of solar radiation data may lead to erroneous solar resource assessment and can lead to project failure [11].

One of the earliest approaches used in the solar radiation data quality analysis is given by, NREL SERI-QC [12]. Here they have introduced some dimensionless components like transmittance coefficients for global, diffuse and direct radiation components. If we plot beam and global coefficients, then we can easily verify the quality of data. Some possible plots for different sky conditions and alignment errors are described in detail later on. Many researchers across the world have contributed to this field: Some have worked on filtering of the GHI and some on combination with DHI and DNI [13]. These GHI, DHI, and DNI data sets are our prime inputs, and the basic approaches, tests, and limits which check the authenticity of data, are discussed in this article. Due to huge variation in data across the world, standard guidelines are established (BSRN) [7]. Tests and guidelines discussed above also involve some meteorological data, which help the data manager to verify the actual problem of disturbed radiation data. The final objective of this research is to provide best class data, which will be free from all errors and shows the real environmental picture of location being analyzed.

2 Database and Site Selection

According to the solar radiation map generated by MNRE and C-WET [4, 14], the north-west region of India is having very good quality of solar radiation and the area is also barren with highest clear sky and sunshine hours (around 300 days). The location selected in this work is Jodhpur (Rajasthan) as many solar installations are expected to come up in this region. The solar radiation data for the month of August, 2015 is selected, as in this month we get the highest variability due to sand-storms, rain and heavy cloud overcast conditions. The detailed description of selected location can be seen in Table 1.

3 Methodology

For ground-based solar radiation data, quality tests to be used are explained below. The tests and procedures follow international practices developed by NREL SERI-QC [12] and WMO BSRN [7]. These standards and tests are elaborated by MESOR [15] and further improved by experience gained by CIEMAT [13], DLR [16] and other ground-based radiation measuring networks. This work is followed by RMIB [17], MNRE C-WET [18] and many more groups around the world. The mathematical approach described in the next section is taken from these sources.

Quality control tests are applied on ground data measured at IMD Jodhpur station.

Table 1 Description of test location

S. No.		
1.	Location	Jodhpur (Rajasthan) Climate: Hot and Dry (MNRE handbook) Latitude/Longitude: 26.25° North/73.04° East
2.	Data Source	IMD, Ground Measurements 10 min (Frequency) High/Medium Maintenance Schedule
3.	Equipment's	Pyranometer (First Class Standard (PSP)—Eppley) Pyrheliometer (Secondary Standard (NIP)—Eppley) Ultrasonic Wind, Pressure, Temperature, Relative Humidity Sensor and Rain Gauge

3.1 Physical Limit Tests [16]

It is the maximum and minimum limit which can be achieved by solar radiation at the most optimal climate conditions. The maximum value which can be achieved is extraterrestrial radiation (1365 W/m^2), but it is affected by the latitude and orientation of the location. The minimum value may go below zero (i.e., −4 W/m^2) due to radiative cooling at night. These limits for various solar radiation components are given below in Eqs. (1)–(3):

$$\mathrm{GHI_{max}} = \mathrm{DNI_0} \times 1.5 \times (\cos(z))^{1.2} + 100, \quad \mathrm{GHI_{min}} = -4\ \left(\mathrm{W/m^2}\right) \tag{1}$$

$$\mathrm{DNI_{max}} = \mathrm{DNI_0}, \quad \mathrm{DNI_{min}} = -4\ \left(\mathrm{W/m^2}\right) \tag{2}$$

$$\mathrm{DHI_{max}} = \mathrm{DNI_0} \times 0.95 \times (\cos(z))^{1.2} + 50, \quad \mathrm{DHI_{min}} = -4\ \left(\mathrm{W/m^2}\right) \tag{3}$$

$$\mathrm{DNI_0} = I_E \times \varepsilon$$

where $\mathrm{GHI_{max}}$, $\mathrm{DNI_{max}}$, $\mathrm{DHI_{max}}$ are maximum theoretical radiation and $\mathrm{GHI_{min}}$, $\mathrm{DNI_{min}}$, $\mathrm{DHI_{min}}$ are minimum theoretical radiation received on selected location.

3.2 Extreme Rare Limit Tests [16]

Here the limit is more refined and these limits are calculated for the most possible radiation values obtained under clear sky conditions in the clear and dry atmosphere (see Eqs. 4–6). The diffuse (Rayleigh) limit is calculated using the work of Long and Shi [19]. Maximum and minimum limits possible for clear sky are given below in terms of the extraterrestrial radiation (ETR):

$$0.03 \times I_E \times \cos(z) < \mathrm{GHI} < \left(1.2 \times I_E \times (\cos(z))^{1.2} + 50\right)\ (\mathrm{W/m^2}) \tag{4}$$

$$0 < \mathrm{DNI} < \left(0.95 \times I_E \times (\cos(z))^{1.2} + 10\right)\ (\mathrm{W/m^2}) \tag{5}$$

$$0.03 \times I_E \times \cos(z) < \mathrm{DHI} < \left(0.75 \times I_E \times (\cos(z))^{1.2} + 30\right)\ (\mathrm{W/m^2}) \tag{6}$$

3.3 Coherence and Correlation Between Radiation Components [16]

The two radiation components (DHI and DNI) are correlated to GHI. Their mutual ratios and correlation with extraterrestrial and theoretically calculated values are also used for data quality analysis. The relations shown in Eqs. (7)–(10) are valid for all types of environmental conditions, and standard limits are calculated. These indicate the quality status of the solar radiation data.

$$\mathrm{DHI/GHI} > 1.05, \quad \text{for } z < 75^\circ \tag{7}$$

$$\mathrm{DHI/GHI} > 1.1, \quad \text{for } 93^\circ > z > 75^\circ \tag{8}$$

$$\left|1 - \frac{\mathrm{GHI}}{\mathrm{DHI} + \mathrm{DNI} \times \cos(z)}\right| > 0.08, \quad \text{for } z < 75^\circ \tag{9}$$

$$\left|1 - \frac{\mathrm{GHI}}{\mathrm{DHI} + \mathrm{DNI} \times \cos(z)}\right| > 0.15, \quad \text{for } 93^\circ > z > 75^\circ \tag{10}$$

3.4 *Tracking Error Test [18]*

For measuring DNI, the device used has to be kept at constant tracking mode, so that equipment is always in direct focus of sun. The angular misalignments leads to tracking error and the effect of this error can be easily seen by making relevant plots of the given time-series. If equipment is not tracking properly then DNI values go down and DHI tends to meet the value of GHI. The change in radiation for DHI and DNI are identified by using Eq. (11) given below:

$$\begin{aligned} &\frac{\mathrm{DHI} + \mathrm{DNI} \times \cos(z)}{\mathrm{GHI_{clear}}} < 0.85, \quad \text{for } z < 75^\circ \quad \mathrm{GHI} > 50\ \left(\mathrm{W/m^2}\right) \\ &\frac{\mathrm{DHI}}{\mathrm{DHI} + \mathrm{DNI} \times \cos(z)} < 0.85, \quad \text{for } z < 75^\circ \quad \mathrm{GHI} > 50\ \left(\mathrm{W/m^2}\right) \end{aligned} \tag{11}$$

where $\mathrm{GHI_{clear}}$ is the GHI on a clear sky day.

3.5 *Visual K-Test [16]*

Radiation Transmittance values are defined as follows: Clearness index (K_t) is the ratio of measured global radiation component with product of extraterrestrial radiation and solar zenith angle. Direct beam transmittance (K_n) is ratio of measured direct radiation component received on horizontal surface with theoretical possible radiation (ETR). Diffuse horizontal transmittance (K) is ratio of measured diffuse radiation component with measured global radiation component. These are calculated by following NREL SERI-QC approach [12]. The location-specific transmittance ratio ranges can be evaluated by either taking standard guidelines of climate types or manual calculation from clear sky days (see Eqs. 12–14). The data produced after this analysis is further visualized and fit in suggested envelops.

$$K_n > K_t, \quad \text{condition for sensor cleaning case}$$
$$K_n (= DNI/I_E) \text{ ratio range } (0-0.8) \tag{12}$$
$$K_t (= GHI/I_E \times cos(z)) \text{ ratio range } (0-1.0)$$

$$K (= DHI/GHI) > 1.05 \text{ for } \quad \text{zenith angle}(z) < 75^\circ \tag{13}$$

$$K (= DHI/GHI) > 1.10 \quad \text{for} \quad \text{zenith angle}(z) > 75^\circ \tag{14}$$

4 Data Analysis

In the selected location (Jodhpur), detailed analysis was done for the month of August, 2015. The available data frequency is 10 min, which is then averaged to hourly values. Only sunshine hours (7 AM to 5 PM) are selected for analysis.

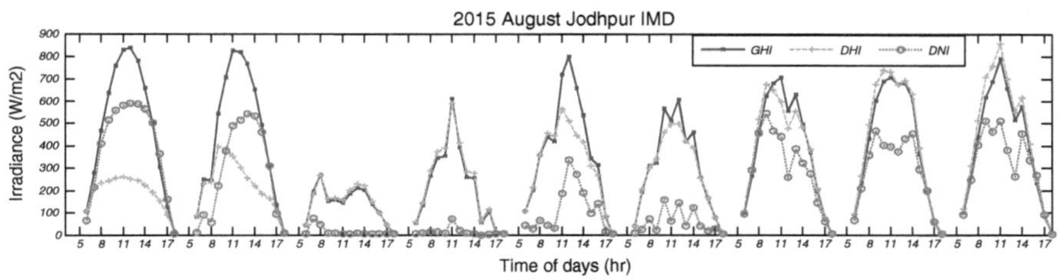

Fig. 1 Time series plot Jodhpur, Aug-2015 (selected days-clear, rainy, cloudy and measurement error, where sunshine duration: 7 AM to 5 PM selected)

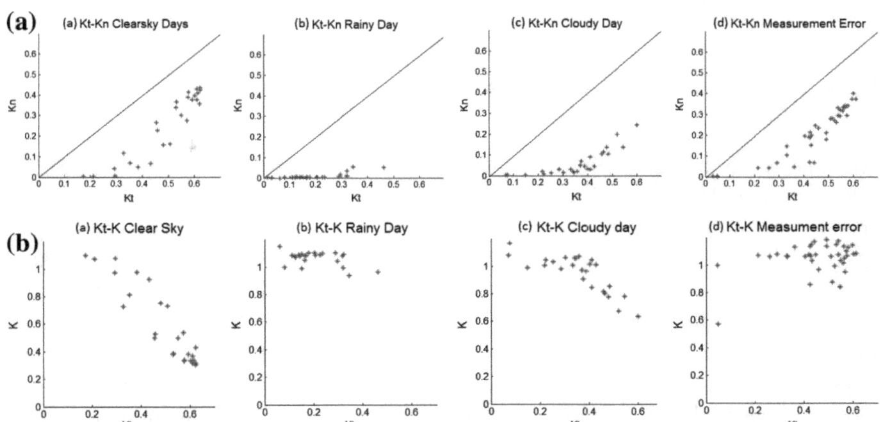

Fig. 2 **a** Transmittance plot (Kt-Kn) for *a* clear, *b* rainy, *c* cloudy and *d* measurement error days, **b** transmittance plot (Kt-K) for *a* clear, *b* rainy, *c* cloudy and *d* measurement error day

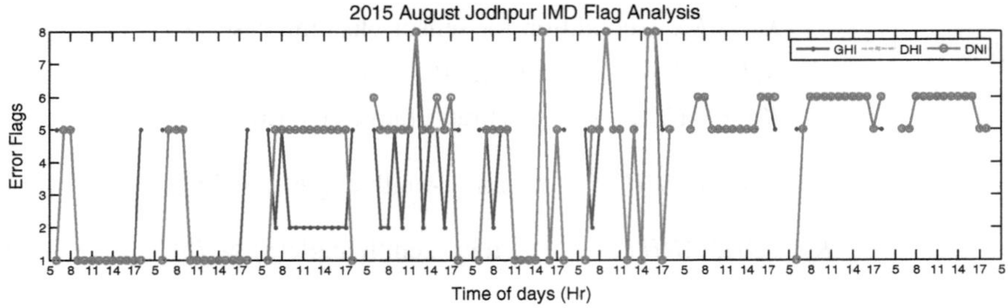

Fig. 3 Associated errors flags for selected days (clear, rainy, cloudy and measurement error)

Table 2 Quality control plot analysis (August 2015 analysis, all results are in %)

	Missing values	Correct values	Min. limit	Max. limit	Clear sky limit	Coherence test	Co-relation	Rayleigh limit	Track error
Flags	0	1	2	3	4	5	6	7	8
DNI	0	36.65	0	1.7	5.8	28.4	7	0	20.2
GHI	0	36	0	0	15.8	25.8	7	0	14.9
DHI	0	37.5	0	2.0	0	31.41	7	0	21.7

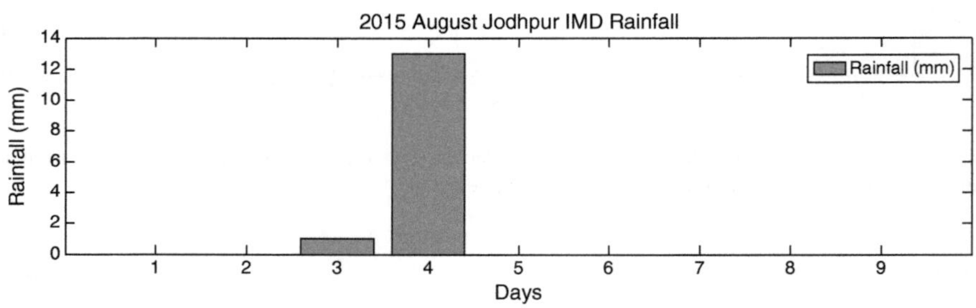

Fig. 4 Rainfall data plot for selected days of August 2015 (IMD Jodhpur)

All quality control guidelines are discussed above and for determining extreme diffuse range, Rayleigh clear sky limit is calculated [19]. Each test is given a specific flag number, and in visual test only basic criteria are checked. The flagging approach is taken from Mitra et al. [18]. Now the complete data quality analysis is applied on the selected data (see Figs. 1, 2 and 3). The results obtained after analysis can easily be understood by referring to Table 2. It may be noted that in Fig. 1, the first 2 days refer to the clear sky, the 3rd and 4th days refer to the rainy days, the 5th and 6th days refer to the cloudy days and the last three days depict the error in measured DHI due to a possible instrument error. The transmittance values shown in Fig. 2, quality test error flags shown in Fig. 3 and rain fall data shown in Fig. 4 also confirm this.

5 Discussion

The results obtained are compared with the rainfall data, taken from IMD and NASA rainfall database (Fig. 4). Based on the detailed visualization of the data plots shown in Figs. 2 and 3 following observations can be made:

(i) On the clear sky days, the data quality is acceptable for further use.
(ii) Rainy days can be seen in month of August. On typical days like this, the usual condition of variable cloud intensity may be present. These are shown as third and fourth days (as rainy days) and 4th and 5th (as cloudy days) in Figs. 1, 2 and 3. By correlating the rain fall data with the solar radiation components during this period further conclusion can be drawn. Unfortunately, cloud cover data was not available for this period. It is suggested that additional weather inputs such as these should be used to identify the source of missing values when standard quality control tests fail.
(iii) For the last 3 days shown in Figs. 1, 2 and 3, the DHI values are constantly increasing, and there is no climatic event present. The reason for this type of response from measuring instrument is probably due to shade-ring correction problem. This can also be seen from the transmittance plots as given in Fig. 2.

The identification of missing values in solar radiation databases and their gap filling is an ongoing area of research. Data visualization can help in identifying various anomalies. If the source of missing values is correlated with additional weather parameters such as rain fall and cloud cover etc., then best-filling procedure can be applied, which satisfies the standard characteristics of that location.

The most important factors due to which measurement error arises are less frequent cleaning and maintenance schedules. So if we clean the equipment frequently (cleaning interval decided according to station location) and the operator is well-acquainted with the system, the majority of the data problems will be solved.

Acknowledgements The authors thank the IMD, Pune for supplying the data of Jodhpur and the MNRE, C-WET personnel (Dr. G. Giridhar and his colleagues) for their help in organizing an SRRA training program at IIT Jodhpur.

References

1. MNRE, Govt. of India. http://www.mnre.gov.in/solar-mission/jnnsm/introduction-2/. Last assessed on 20 Mar 16
2. Press Information Bureau, Govt. of India. http://pib.nic.in/newsite/pmreleases.aspx?mincode=28. Last assessed on 20 Mar 16
3. IMD Data Center. http://www.imd.gov.in/Welcome%20To%20IMD/Welcome.php. Last assessed on 20 Mar 16
4. MNRE, Govt. of India. http://mnre.gov.in/centers/about-sec-2/hand-book-on-energy-conscious-buildings/. Last assessed on 20 Mar 16

5. N.K. Bansal, G. Minke, *Climatic Zones and Rural Housing in India* (Kernforschungsanlage, Juelich, Germany, 1988)
6. SP 7:2005, *National Building Code of India 2005* (Bureau of Indian Standards, New Delhi, 2005)
7. WMO, *CIMO Guide to Meteorological Instruments and Methods of Observation, Preliminary Seventh Edition, WMO-No. 8* (Secretariat of the World Meteorological Organization, Geneva, Switzerland, 2006)
8. ISO 9060, 1990, *Solar Energy—Specification and Classification of Instruments for Measuring Hemispherical Solar and Direct Solar Radiation*. International Organization for Standardization, 01 Nov 2008
9. S. Younes, R. Claywell, T. Muneer, Quality control of solar radiation data: present status and proposed new approaches. Energy **30**, 1533–1549 (2005)
10. N. Geuder, V. Quaschning, Soiling of irradiation sensors and methods for soiling correction. Sol. Energy **80**(11), 1402–1409 (2006)
11. Ratings Fitch, in *Rating Criteria for Solar Power Projects*. Utility-Scale Photovoltaic, Concentrating Photovoltaic, and Concentrating Solar Power. 2013. Global Infrastructure & Project Finance. www.fitchratings.com/site/criteria/infrastructure
12. E. Maxwell, S. Wilcox, M. Rymes, User's manual for SERI QC software, assessing the quality of solar radiation data. Report no. NREL-TP-463-5608. 1617 Cole Boulevard, Golden, National Renewable Energy Laboratory, Colorado (1993)
13. S. Moreno-Tejera, L. Ramírez-Santigosa, M.A. Silva-Pérez, A proposed methodology for quick assessment of timestamp and quality control results of solar radiation data. Renew. Energy **78**, 531–537 (2015)
14. Indian Solar Radiation Atlas. http://niwe.res.in/assets/Docu/srra_solar_radiation_brochure.pdf. Last assessed on 20 Mar 16
15. C. Hoyer-Klick, D. Dumortier, A. Tsvetkov, J. Polo, J.L. Torres, C. Kurz, P. Ineichen, *MESOR Existing Ground Data Sets, D 1.1.2* (2009)
16. N. Geuder, F. Wolferertstetter, S. Wilbert, D. Schűler, R. Affolter, B. Kraas, E. Lűpfert, B. Espinar, Screening and flagging of solar irradiation and ancillary meteorological data. Energy Procedia **69**, 1989–1998 (2015)
17. M. Journée, C. Bertrand, Quality control of solar radiation data within the RMIB solar measurements network. Sol. Energy **85**, 72–86 (2011)
18. I. Mitra, K. Chhatbar, A. Kumar, G. Girdhar, R. Vashistha, R. Meyer, M. Schwandt, Solmap: project in India's solar resource assesment. Int. J. Renew. Energy Dev. **3**(3), 207–216 (2014)
19. C.H. Long, Y. Shi, An automated quality assessment and control algorithm for surface radiation measurements. Open Atmos. Sci. J. **2**, 23–27 (2008)

Design of Field Layout for Central Receiver System to Generate 100–150 kW Solar Thermal Power

Pedamallu V.V.N.S.P. Raju and V. Narayanan

Abstract In harvesting Solar Energy via, solar thermal power, design of optical field layout is one of the important parameter for efficient land usage and also optimizing the field for energy efficiency available in a given region. Thus, in concentrated solar power (CSP) we investigate a procedure for the design and optimization of heliostat field layout for a thermal power 100–150 kW having 5–7 operational hours. First, in the design of heliostat for a given power requirement, we propose the position of the heliostat, along with curvature and dimension of the heliostat. Second, for the optimal layout, shadow lengths, altitude, and azimuthal angles were considered along with different seasons at different times with respect to Jodhpur location latitude and longitude angles. The position of the heliostat were considered, which gives minimum cosine losses. The total number of heliostat considered were ranging from 50 to 60, having varying curvature depending on its distance from the receiver, were used in the layout design. Each mirror having 6.25 m^2 area and are approximately arranged in the field of 0.57 acres area of land. The field layouts are discussed in details in terms of blocking and shadowing effects. Finally, a proposed layout is discussed for the application of CSP.

Keywords Thermal power · Field layout · Heliostat · Tower · Central receiver · Efficiency

P.V.V.N.S.P. Raju (✉)
Centre for Solar Energy Technology, IIT Jodhpur, Jodhpur, India
e-mail: rajeshpedamallu@gmail.com

V. Narayanan
Department of Physics, IIT Jodhpur, Jodhpur, India
e-mail: vnara@iitj.ac.in

L. Chandra and A. Dixit (eds.), *Concentrated Solar Thermal Energy Technologies*, Springer Proceedings in Energy, https://doi.org/10.1007/978-981-10-4576-9_5

1 Introduction

Large-scale solar fields are used to collect the solar radiation with help of reflectors and concentrate the radiation to attain high temperatures in the receiver. A suitable mechanism is employed to convert the available high temperature at the receiver to produce electricity [1]. Among the various types of concentrating the solar radiation, the Central receivers with heliostats (mirrors) are regularly deployed [2]. In central receiver systems, high concentration of solar irradiation can be achieved thus resulting in extremely high temperature. These systems are characterized for large power levels (1–500 MW) and high temperatures (500–800 °C). Central receiver systems are based on a field layout of individual two axes Sun-tracking mirrors (heliostats). Mirrors reflect the incident solar radiation to the receiver at the top of a centrally located tower. Generally, central receiver systems are made up of heliostats, towers, receivers, heat transfer devices, thermal storage devices, and power generation parts. Further, design of solar filed depends on power requirement with available radiation in the given region. In this work, the central receiver system field layout has been designed for Jodhpur location having annual average of Direct Normal Irradiance (DNI) in Rajasthan (Jodhpur) is approximately 5.5 kWh per square meter per day [3, 4]. Accordingly, the proposed 150 kW thermal power for the Jodhpur location, we suggest the size of each heliostat having area of 6.25 m^2 with 50–60 heliostats can generate suggested power requirements with optimal land use.

2 Optical Field Efficiency

The performance of the heliostat field defines in terms of optical efficiency. It is defined as the ratio of the net power intercepted by the receiver to the power incident normally on the field [5]. Various types of losses affect the optical efficiency. The types of loss are as follows: mirror losses, atmospheric losses, spillage losses, cosine losses, shadowing, and blocking losses; which reduce the optical efficiency of the heliostat field. The optical efficiency of the field can be given by [6]:

$$\eta_{\text{field}} = \eta_{\text{mirror}} \times \eta_{\text{atm}} \times \eta_{\text{spillage}} \times \eta_{\text{cosine}} \times \eta_{\text{S\&B}} \tag{1}$$

where η_{mirror} is the efficiency of mirror reflectivity, η_{atm} is atmospheric efficiency, η_{spillage} is spillage efficiency, η_{cosine} is cosine efficiency and $\eta_{\text{S\&B}}$ is shadowing and blocking efficiency. These efficiencies are explained in the proceeding section.

2.1 *Efficiency of Mirror η_{mirror}*

Efficiency of the mirror depends upon the reflectivity of the heliostat mirrors. It completely depends on which type of material coated on backside of the mirror.

Heliostat mirror surfaces reflect most of the incident radiation but it also absorbs a portion of energy. Currently, 80–95% reflectivity mirrors are commercially available [7] and a standard value for the reflectivity 88% is considered in this field design [8].

2.2 *Efficiency of Atmosphere η_{atm}*

A portion of the solar radiation of the reflected rays are scattered and absorbed by the atmosphere and all the reflected solar radiation would not reach the receiver. This loss is referred as atmospheric attenuation loss. The atmospheric attenuation loss is a function of distance d_{HM} between each heliostat mirror and the receiver. The atmospheric efficiency can be expressed as [9]:

When $d_{HM} \leq 1000$ m,

$$\eta_{\text{atm}} = 0.99321 - 1.176 \times 10^{-4} d_{HM} + 1.97 \times 10^{-8} d_{HM}^2 \tag{2}$$

When $d_{HM} > 1000$ m,

$$\eta_{\text{atm}} = \mathrm{e}^{-0.0001106 \times d_{HM}} \tag{3}$$

2.3 *Spillage Efficiency $\eta_{spillage}$*

All the portion of the reflected radiation cannot hit the receiver, their by causing Spillage (e.g., parameters like curvature errors, tracking errors, receiver size, etc., influence the spillage). The reflected light rays are limited due to the finite size of the receiver. The size of the aperture of the receiver is minimized to reduce the convection and radiation losses without blocking out too much of solar flux arriving at the receiver. The aperture size of the receiver is typically the same size as the Sun's reflected image from the farthest heliostat to reduce the spillage losses.

2.3.1 Curvature and Size of the Reflected Image of Concave Mirror

The radius of curvature of the mirror '*R*' is the radius of the mirror that forms a complete sphere and the focal length of the mirror '*f*' is the distance between the center of the mirror and the point at which the reflected light meets the principle axis of the mirror.

From Fig. 1, *AB* is incident light ray and *BF* is reflected light ray. *BC* represents the radius of the mirror. The line *BC* bisects the angle *ABF* and *ΔBCF* forms an isosceles triangle. Therefore, the line *BF* and *FC* are equal. We also know that *BF* and *OF* are equal for lenses that are small.

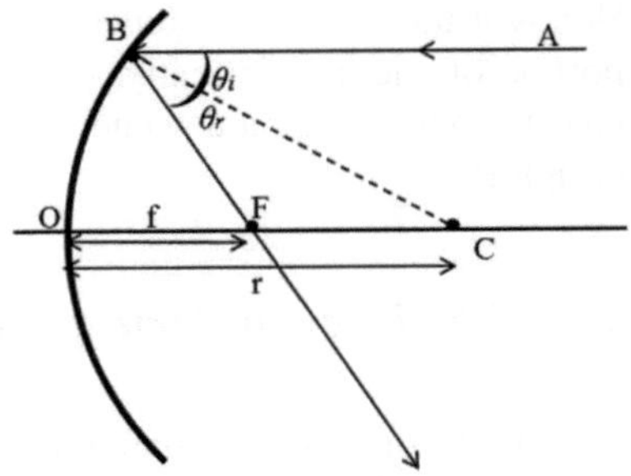

Fig. 1 Relationship between f and r for a concave mirror

$$\begin{aligned} &\therefore BF = FC = OF \; \& \\ &OC = OF + FC \\ &\Rightarrow OC = OF + OF \\ &\Rightarrow OC = 20F \Rightarrow 2F \end{aligned} \tag{4}$$

From above equations, we can say that radius of curvature of the concave mirror $OC(r)$ is equal to twice the focal length of the mirror.

$$r = 2f \tag{5}$$

Each point on a mirror will reflect a cone of rays that matches the angular distribution of the solar source (half-angle of solar rays divergence θ_s = 4.65 milli radian) [2]. The rays from the rim of the mirror will forms the widest spot on the flat target placed at the focal plane of the mirror.

From Fig. 2, the spot size 'd' can be written as

$$d = \frac{2p \sin \theta_s}{\cos \theta_R} \tag{6}$$

where 'a' is the aperture size of the mirror. When received incidence angle θ_R is small p is approximately equal to f, then image spot size being

$$d = \frac{2f \sin \theta_s}{\cos 0} = 2f \sin \theta_s \tag{7}$$

't' is the thickness or sag of the mirror. From the Sag equation we can measure the thickness 't' as

$$t = r - \sqrt{r^2 - (a/2)^2} \tag{8}$$

where 'r' is the radius of curvature of the mirror.

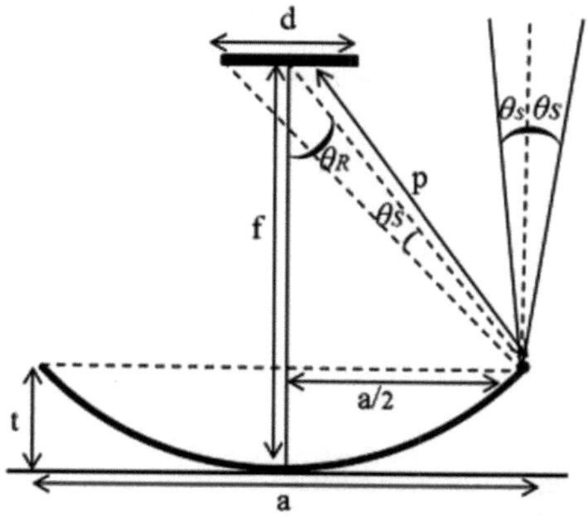

Fig. 2 Spot size of a heliostat mirror

2.4 *Cosine Efficiency* η_{cosine}

The cosine efficiency depends on both Sun's position and location of the individual heliostat mirrors with respect to the receiver. It is related to the cosine angle between the incident Sun ray vector and normal vector of the heliostat mirror [10]. The effective reflected area of the mirror related by cosine angle. Larger cosine angles reduce the effective reflection area.

Heliostat A has small cosine losses compared to heliostat B as shown in Fig. 3. The efficient performance of the heliostats is achieved by keeping heliostats in north side of the tower, wherein heliostats are opposite to sun. During morning hours, North West field heliostats with respect to the tower having more efficiency compared to North East field heliostats and evening time it is opposite to the field. The cosine angle between the Sun vector and the reflected vector from mirror to receiver 2θ can measure by using Sun and Earth geometry and the cosine angle 'θ' represents the cosine efficiency of the Heliostat mirror (Fig. 4).

$$\cos 2\theta = S.r \tag{9}$$

where 'S' is sun ray vector and 'r' is the reflected ray unit vector from mirror to receiver.

Sun unit vector expressed in terms of altitude and azimuth angle as [11]:

$$S = S_z i + S_e j + S_n k \tag{10}$$

where

$$\begin{aligned} S_z &= \sin\alpha \\ S_e &= \cos\alpha \sin A \\ S_n &= \cos\alpha \cos A \end{aligned} \tag{11}$$

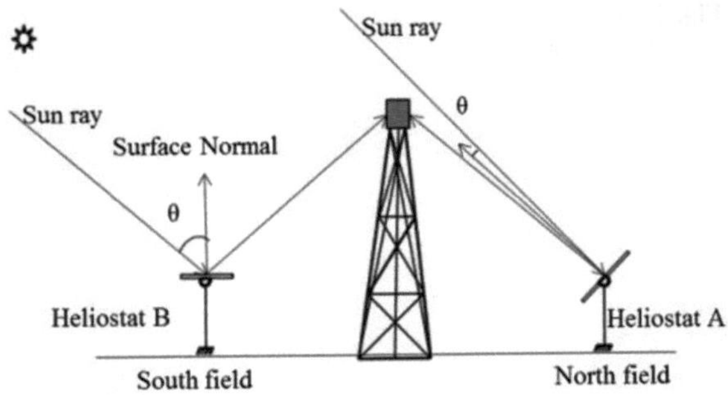

Fig. 3 The cosine effect for two heliostats in opposite directions from the tower

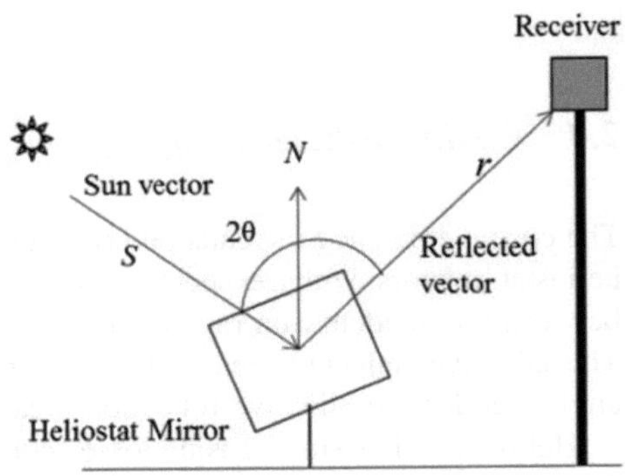

Fig. 4 The cosine angle between the heliostat to the receiver

α is altitude angle and A is the azimuth angle. Reflected ray unit vector can be written as

$$r = \frac{(x_r - x_m)i + (y_r - y_m)j + (z_r - z_m)k}{\sqrt{(x_r - x_m)^2 + (y_r - y_m)^2 + (z_r - z_m)^2}} \tag{12}$$

2.5 Shadowing and Blocking Efficiency $\eta_{S\&B}$

The reflected radiation from heliostat mirrors are blocked by adjacent heliostats while concentrating on to the receiver and incoming solar radiation is obstructed by the neighboring heliostats cause the shadowing effect. Shadowing and blocking effects plays a crucial role in optimized field lay out design. These effects reduce the solar radiation reaching the receiver. The amount of shadowing and blocking is a function of Sun's position angles, tower height, and heliostat mirror location in the field. When a heliostat is placed in between the two heliostats in the adjacent row, the reflected light rays can pass between the adjacent heliostats on the way to the receiver with minimal blocking effect. Shadowing and blocking efficiency can maximize by increasing the separation between heliostats, but it consumes the more land area. We maximize the efficiency by using minimum land area in such a way that the minimum distance between two adjacent heliostats equal to the characteristic diameter (DM) of each heliostat, this is equal to the diagonal of the heliostat plus the separation distance, i.e., [5]

$$\mathrm{DM} = \mathrm{LH}\left(\sqrt{1+f^2} + ds\right) \tag{13}$$

where 'f' is the heliostat width to length ration and 'ds' is the ratio of heliostat separation distance to heliostat length (Fig. 5).

For no blocking, the minimum value of 'ds' and 'DM' as given by Collado and Turegano [12] is

$$ds_{\min} = 2f - \sqrt{1+f^2} \tag{14}$$

Therefore,

$$\mathrm{DM}_{\min} = 2\mathrm{WH} \tag{15}$$

Fig. 5 Minimum distance between adjacent heliostats

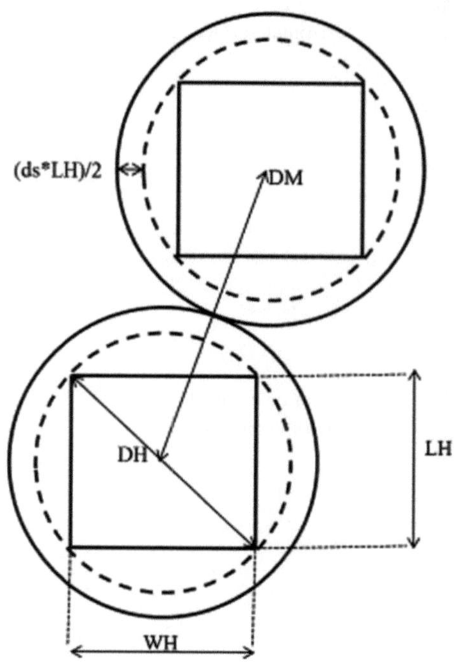

Knowing of Sun's position is very important in predicting the length and direction of a shadow of the heliostat. Since Sun's light travels in straight line, the projected shadows on the ground can be measured using simple geometry.

Figure 6 shows a heliostat on a horizontal surface and 'α' is the altitude angle of the Sun's rays making with the horizontal surface. In this case, the length of the shadow can be expressed as:

$$L = h/\tan\alpha \tag{16}$$

where 'h' is the height of the mirror and Fig. 7 gives the shadow length in case of inclined plane. It can be expressed as [13]:

$$A_h = \sin\beta\cot\alpha - \cos\beta \tag{17}$$

where 'β' is the inclined plane angle and 'α' is altitude angle of Sun's rays making with horizontal surface. Altitude angle and azimuth angle gives the length and direction of the shadow.

The optical efficiency of the heliostat caused by shadowing and blocking effect are computed using ray-tracing technique. The reflective area of the test heliostat involved in the shadowing and blocking as represented in Figs. 8 and 9, respectively.

Fig. 6 Projection of vertical plane on a horizontal surface

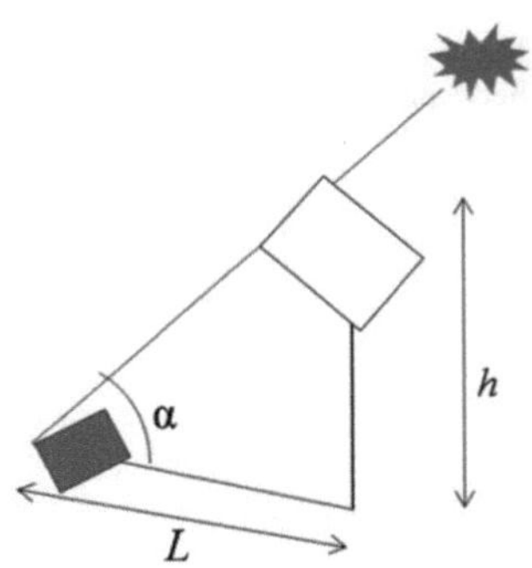

Fig. 7 Projection of inclined plane on a horizontal surface

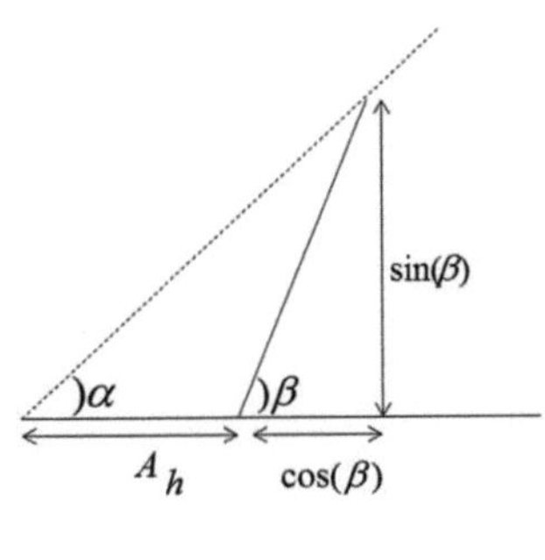

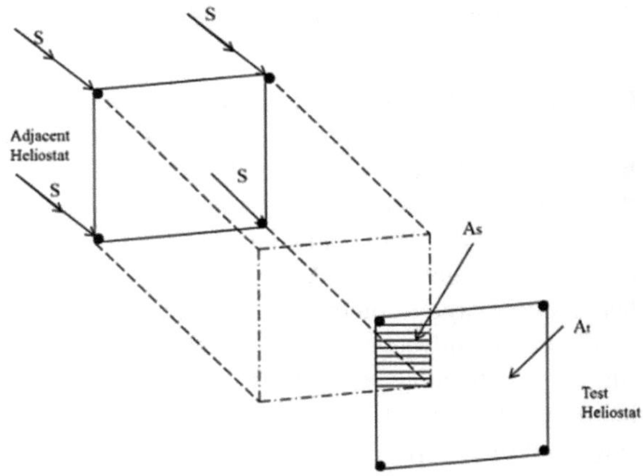

Fig. 8 Shadowing effect

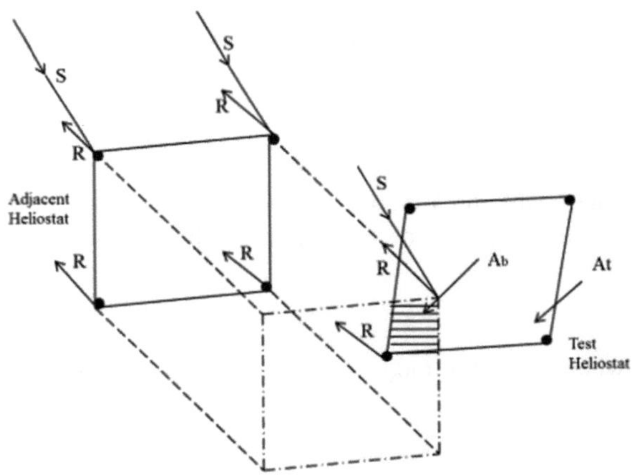

Fig. 9 Blocking effect

'A_t' is the area of the Test heliostat, 'A_s' is the shadowing area caused by the adjacent heliostat and 'A_b' is blocking area on Test heliostat due to blocking of some reflected sun rays by the adjacent heliostat. The shadowing efficiency and blocking efficiency can be calculated using below relations after finding the shadowing and blocking areas using ray-tracing techniques [14].

$$\eta_{\text{Shadow}} = \frac{(A_t - A_s)}{A_t} \tag{18}$$

$$\eta_{\text{Block}} = \frac{(A_t - A_b)}{A_t} \tag{19}$$

3 Optimized Field Layout Design

Closely packed heliostat field exhibits more shadowing and blocking losses. In general, a radial stager pattern [15] is used to minimize the land usage as well as blocking and shadowing losses. The heliostats are closely packed near the tower but those heliostats should be separated to prevent mechanical interference by the adjacent mirrors. The azimuthal spacing increases when heliostats located farther distance from the tower. Additional heliostats are added when azimuthal spacing become too large in outer rings and a new stager pattern is established.

Figure 10 shows an optimized field layout design. In this pattern, two types of rings are defined as essential rings and staggered rings. The rings which have heliostats on the north axis are called essential rings (E_1, E_2, E_3, E_4, E_5, E_6) while in staggered rings (S_1, S_2, S_3, S_4, S_5), the rings which don't have heliostats on the north axis in the field [6]. The distance between the essential or staggered rings called radial spacing 'ΔR' and the separation between heliostat in the same rings is called azimuthal spacing 'ΔA'. These two parameters defined as [15]:

$$\Delta R = HM(1.44\cot\theta_{\mathrm{L}} - 1.094 + 3.068\theta_{\mathrm{L}} - 1.1256\theta_{\mathrm{L}}^2) \tag{20}$$

$$\Delta A = \mathrm{WM}(1.749 + 0.6396\theta_{\mathrm{L}}) + \frac{0.2873}{\theta_{\mathrm{L}} - 0.04902} \tag{21}$$

where HM and WM are the height and width of the heliostat, and the angle θ_{L} is the altitude angle to the receiver from the heliostat location and defined as [15]:

$$\theta_{\mathrm{L}} = \tan^{-1}(1/R_0) \tag{22}$$

where R_0 is the radius of the first essential ring in terms of the tower height H_{T}.

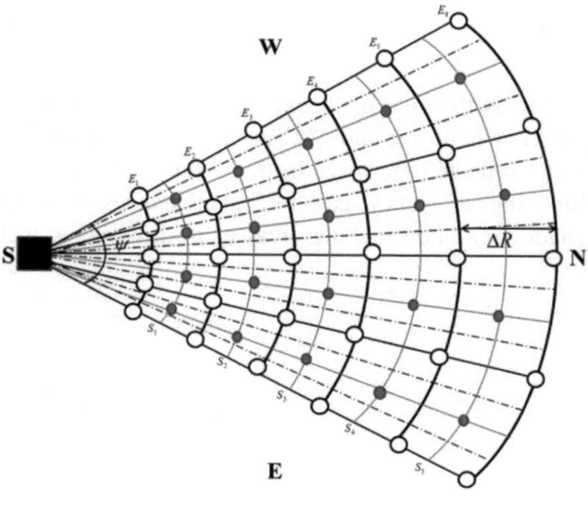

Fig. 10 The radial stagger heliostat layout pattern

According to this azimuthal distance five heliostats kept in the first ring and in the same angular direction 25 heliostats are placed in the next five essential rings (*E*). In between six essential rings, five staggered rings are there and each ring contains four heliostats in between two the adjacent essential rings. In this manner, 50 heliostats are placed in the field. Some additional heliostats can be added in between these heliostats where azimuthal spacing becoming too large in the outer rings.

4 Analysis of the Heliostat Field

The designed optimized field layout is shown in Fig. 10. The proposed height and width of the heliostat is 2.5×2.5 m^2 is chosen and heliostats kept at 1.5 m height from the surface of the ground (HM equal to 4 m height) minimizing the dust deposition on the mirrors. Height of the tower H_T is taken 18 m and the first essential ring radius R_0 kept as equal height of the tower. The field occupies 0.57 m^2 acres land area with the rim angle of 76.63° (field view of the receiver aperture). The required field area is calculated with different radius of essential ring R_0 and height of the tower H_0 and measured radial spacing, azimuthal spacing, rim angle of the field, and occupied field areas are mentioned in Table 1.

From the above tabular column, the increment in the rim angle ψ (field of view of the receiver aperture) and field area with decrement in the radius of the first essential ring R_0 is observed. In literature, it is reported cavity receivers or volumetric receivers are limited by an acceptance angle 60–120° [16]. Therefore, multiple cavities are placed adjacent to each other or the heliostat field is limited to the view of the cavity aperture acceptance angle [16]. When the radius of the first essential ring R_0 is 10 m, field of view (rim angle ψ) 146° exceeding the limit of acceptance angle

Table 1 The calculated field area at different radius of essential ring R_0 and different height of the tower H_0

S. No.	Height of the tower H_0 (m)	Radius of the first essential ring R_0 (m)	Radial spacing ΔR (m)	Azimuthal spacing for first ring ΔA_1 (m)	Rim angle of the field ψ (°)	Area of the field (m^2)	Field in terms of acres
1	18	10	6.78	6.36	146	2449	0.6
2	18	13.5	7.45	6.18	105	2358	0.583
3	18	15	7.72	6.12	93.56	2343.9	0.58
4	18	18	8.25	6.02	76.63	2345	0.57
5	20	18	7.93	6.07	77.37	2243.98	0.56
6	25	18	7.35	6.206	79	2068.68	0.511
7	30	18	6.94	6.31	80.37	1949.79	0.482
8	35	18	6.63	6.4	81.47	1859.34	0.4595
9	20	20	8.25	6.02	68.96	2256.07	0.56
10	30	30	8.25	6.02	45.98	2035.48	0.503

Table 2 Specifications of heliostat field and optical field efficiency

Parameters	Specifications
Total number of heliostats Height of the tower H_0	50 18
Radius of the first essential ring R_0	18 m
Total reflective area of the heliostat field	312.5 m^2
Aperture size of the receiver	0.57 m
Mirror efficiency	0.88
Atmospheric efficiency	0.9887
Spillage efficiency (in the absence of tracking and curvature errors of Heliostat mirrors)	1
Shadowing and blocking efficiency	0.92
Cosine efficiency	0.9104
Total field optical efficiency	0.7287

thus not suited for above said receiver. While minimum radius R_0, equal to height of the tower H_0 (First row) is inside the limit of acceptance angle and occupying minimal land usage. When height of the tower H_0 is increasing, this condition satisfying limit of acceptance angle and occupying minimum land area but due to the decrement in the radial spacing ΔR, shadowing and blocking effects are increasing and field having poor optical field efficiency. Further, when R_0 and H_0 are increasing in such a way that both the ratio should equal to 1, in this case field satisfying no blocking and shadowing conditions and occupying the minimum land area but due to the increment the in distance between mirrors to receiver, cosine losses and atmospheric losses are increasing slightly. The combination of tower height H_0 and first essential ring radius R_0 18 m is giving minimal land area and good optical efficiency. This field configuration satisfying the no blocking condition and having great shadowing, blocking efficiency and cosine efficiency. Shadowing and blocking efficiency is measured using ray-tracing techniques and cosine and atmospheric efficiencies are measured for each heliostat mirror using the relations (10 and 2). The total field optical efficiency (for 50 heliostat mirrors as shown in Fig. 10) is measured for Jodhpur location in India at the time of 10.30 am on 22 June. The optical field efficiency and some important parameters for the field mentioned in Table 2.

Thus, the proposed field layout with minimal area usage can operate 6–7 working hours with highly optical efficiency to generate 100–150 kW thermal power.

5 Conclusions

The optimized solar field design for heliostat and central tower has been investigated. The proposed heliostat field design provides a valuable input for estimation of energy efficiency and optimal land usage along with larger duration of operation in a day. The design of heliostat along with curvature and dimension of the heliostat for a given optical power requirements has been investigated by considering the optimal layout, shadow lengths, altitude and azimuthal angles for given location (Jodhpur). Further, by positioning the heliostat, we find our design which gives minimum cosine losses and also, the effect with different seasons at different times for given location is investigated by location's latitude and longitude angles. A procedure for the design and optimization of heliostat field layout for a thermal power 100–150 kW having 5–7 operational hours has been reported with minimal land usage. Finally, a proposed layout is discussed for the application of CSP.

Acknowledgements The authors are thankful to MNRE and Indian Institute of Technology Jodhpur for their encouragement and financial support.

References

1. M.A. Mustafa, S. Abdelhady, A.A. Elweteedy, Analytical study of an innovated solar power tower (PS10) in Aswan (Mechanical Power Engineering Department, M.T.C., Cairo, Egypt). Int. J. Energy Eng. **2**, 273–278 (2012)
2. K. Lovegrove, W. Stein, *Concentrating Solar Power Technology, Principles, Developments and Applications* (Woodhead Publishing Series in Energy, 2012)
3. A. Kumarankandath, N. Goswami, *The State of Concentrated Solar Power in India* (Centre for Science and Environment, 2015)
4. P.R. Arora, A vital role of concentrating solar power plants of Rajasthan in future electricity demand of India (Research Scholar). Int. J. Sci. Res. Publ. **3**(6) (2013). ISSN 2250-3153
5. F.M.F. Siala, M.E. Elayeb, Mathematical formulation of a graphical method for a no-blocking heliostat field layout (Center For Solar Energy Studies, P.O. Box 12932, Tripoli, Libya). Renew. Energy **23**, 77–92 (2001)
6. Y. Zhou, Y. Zhao, in Heliostat Field Layout Design Solar Tower Plant Based on GPU. Department of Control Science and Engineering, Zhejiang University, China. *19th IFAC World Congress Cape Town, South Africa*, 24–29 Aug 2014
7. M. DiGrazia, G. Jorgensen, *Reflectech Mirror Film: Design Flexibility and Durability in Reflecting Solar Applications*. ReflecTech, Inc. Arvada, CO 80007 USA and National Renewable Energy Laboratory (2010)
8. S.A. Rafeq, Z.M. Zulfattah, A.M. Najib, M.Z.M. Rody, S. Fadhli, M.F.B. Abdollah, M.H.M. Hafidzal, Preliminary study of CST in Malaysia based on field optical efficiency (Faculty of Mechanical Engineering, Universiti Teknikal Malaysia Melaka, 76100 Durian Tunggal, Melaka, Malaysia, ScienceDirect). Procedia Eng. **68**, 238–244 (2013)
9. M. Ewert, O. Navarro Fuentes, *Modelling and Simulation of a Solar Tower Power Plant*. Master Students of Computer Science at RWTH Aachen University, Aachen, Germany (2015)
10. F. Eddhibi, M. Ben Amara, M. Balghouthi, A. Guizani, Optical study of solar tower power plant (Thermal Process Laboratory, Research and Technology center of Energy P.B $N^0$95 2050, Hammam Lif-Tunisia). J. Phys. Conf. Ser. (IOP Publishing Ltd) **596**, 012018 (2015)
11. H.M. Woolf, *On the Computation of Solar Evaluation Angles and the Determination of Sunrise and Sunset Times*. National Aeronautics and Space Administration Report NASA TM-X-164, September (1968)
12. F.J. Collado, J.A. Turegano, Calculation of the annual thermal energy supplied by a defined heliostat Field. Sol. Energy **42**, 65–149 (1989)

13. D. Baldocchi, *Lecture 7, Solar Radiation, Part 3, Earth-Sun Geometry*. Department of Environmental Science, Policy and Management, University of California, Berkeley CA94720, 10 September 2012
14. K.K. Chong, M.H. Tan, Comparison study of two different sun-tracking methods in optical efficiency of heliostat field (Faculty of Engineering and Science, Universiti Tybjy Abdul Rahman, Off Jalan Genting Kelang, Setapak, 53300 Kuala Lumpur, Malaysia). Int. J. Photoenergy **2012** (2012)
15. T.A. Dellin, M.J. Fish, C.L. Yang, *A User's Manual for DELSOL2: A Computer Code for Calculating the Optical Performance and Optimal System Design for Solar Thermal Central Receiver Plants*. Sandia National Labs Report SAND81-8237, August (1981)
16. K.W. Battleson, *Solar Power Tower Design Guide: Solar Thermal Central Receiver Power Systems, A Source of Electricity and/or Process Heat*. Sandia National Labs Report SAND81-8005, April (1981)

Design and Investigation of Parabolic Trough Solar Concentrator

Deepika Tamta and R.P. Saini

Abstract There is a huge potential of solar energy available in India. For converting solar energy into heat and thermal energy, two technological routes are possible. Solar photovoltaic and solar thermal routes are viable to harness solar energy. Solar thermal power are based on the principle of focussing suns energy and transforming it into high temperature heat for producing steam which can further be beneficial in generating electricity using conventional power cycles. Based on literature survey, it has been found that parabolic trough technology is the most promising and low-cost technology. In this study, an attempt has been made to carry out an optical analysis for sizing of parabolic trough concentrator. Further performance of the designed parabolic trough with hot water generation system has been investigated experimentlly. The maximum value of performance parameter is observed around noon, when the solar beam radiation is its maximum. It has been observed that variation in solar radiation and wind velocity influence average receiver temperature. The study may be useful for future studies in order to investigate the effect of other parameters on performance of parabolic trough collectors.

Keywords Solar thermal · Solar photovoltaic · Solar power · Parabolic trough solar concentrator · Conventional power cycles

---

D. Tamta (✉) · R.P. Saini
Alternate Hydro Energy Centre, Indian Institute of Technology Roorkee, Roorkee 247667, Uttarakhand, India
e-mail: tamtadeepika1@gmail.com

R.P. Saini
e-mail: rajsafah@iitr.ac.in

L. Chandra and A. Dixit (eds.), *Concentrated Solar Thermal Energy Technologies*, Springer Proceedings in Energy, https://doi.org/10.1007/978-981-10-4576-9_6

1 Introduction

Solar energy is a renewable source that is inexhaustible and is locally available. Among all the renewable energy sources solar energy has the most potential. For generating electric power from the solar energy, two paths are available namely concentrating solar power and solar photovoltaic. Among them, Solar Thermal Power technology is advantageous because of its less design cost and accessibility of components. There are various solar thermal power technologies that have evolved and developed. Among all, concentrated solar power technology parabolic through solar power technology is most suitable for air water heating and power generation. Solar power technologies works on the principle of focusing solar radiation into small area to produce steam or hot air which can be used for electricity generation using conventional power cycles. Various promising development in the field of parabolic trough concentrator for various application of thermal energy was presented by Fernandez et al. [1]. A broad survey on different solar thermal collector and application was carried out by Kalogirou [2]. Intercept Factor and tracking mode are the major factor for obtaining maximum possible collector efficiency [3].

Numerous studies have been conducted on design and performance aspects of parabolic trough concentrator. Mohamed [4] has done experimental study of parabolic trough under local climate condition. Thomas and Guven [5] studied the design aspects and various techniques to enhance the parabolic trough performance. The collector performance using three heat transfer fluid has been investigated by Islam et al. [6]. Optimum tilt angle for parabolic trough collector with north-south tilt angle has been found by Zhan et al. [7]. The collector's thermal efficiency mainly depends to meteorological conditions such as solar radiation, the room temperature and the cloudiness [8]. For minimizing the cost involved in water heating parabolic trough is most suitable option during winter [9]. Sulaiman et al. [10] investigated the process that is required to examine the cylindrical parabolic trough concentrator performance. Thermal analysis of parabolic trough has been done to calculate important parameter by Kumar et al. [11].

2 Parabolic Trough Solar Concentrator

Parabolic trough concentrator consists a reflective material sheet of parabolic shape when parabola is directed toward sun parallel sun rays falls on collector are reflected onto the absorber tube. The heat transfer fluid which is flowing through absorber tube gets heated to very high temperature by reflected solar beam radiation. For household and industrial works this heated fluid can be used. Parabolic trough concentrator can effectively produce temperature in a range of 50 and 400 °C. The orientation of parabolic trough depends upon the season.

The design of the parabolic trough solar concentrator is done using PARABOLA CACULATER 2.0 software. The initial specification for Parabolic Trough collector design is obtained by considering parabolic equation:

$$x^2 = 4aY$$

where

Y Distance along vertical axis
a Focal length
X Distance along horizontal axis.

3 Design of Parabolic Trough Concentrator

For designing the parabolic trough rim angle, rim radius, aperture width, focal length, and length of parabola are basic parameters. Rim angle plays most important role for designing of parabolic concentrator. The schematic of parabolic trough showing all parameters is shown in Fig. 1.

Stainless steel SS304 has been used as reflecting surface. The sheet was curved to form parabolic trough of 1.25 m of length and 1.74 m of aperture width. The aperture area is 2.18 m^2. A simple parabolic equation is used for designing the parabolic trough concentrator.

$$Y = ax^2$$

where $a = \frac{1}{4f}$

$$\frac{f}{W} = 0.25 * \cot\frac{\varphi_r}{2}$$

$$r_r = \frac{2f}{1 + \cos\varphi_r}$$

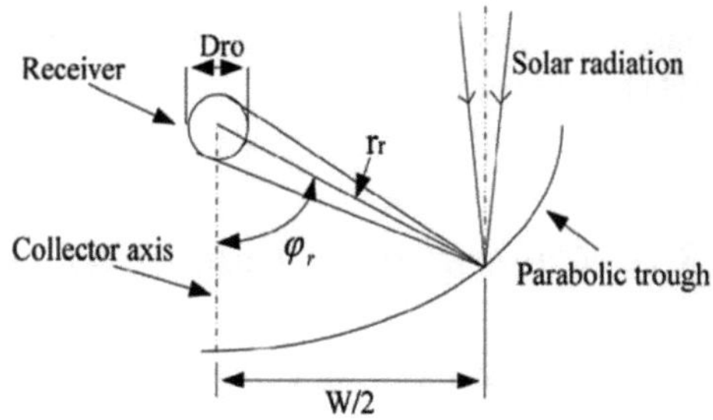

Fig. 1 Schematic of a parabolic trough concentrator

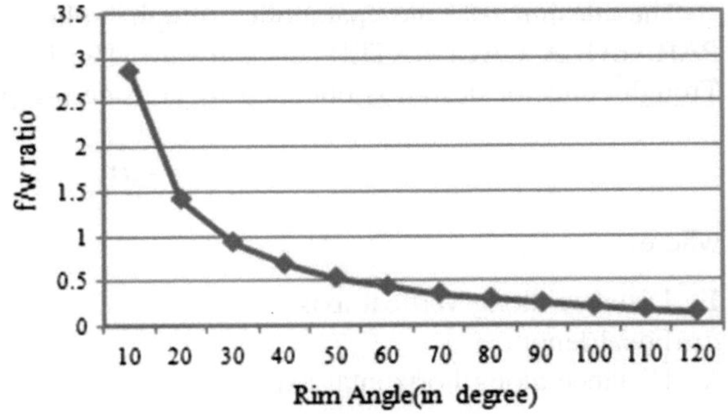

Fig. 2 Focus aperture ratio versus rim angle

where f, W, r_r and φ_r shows focal length, aperture width, rim radius, and rim angle, respectively, Fig. 2 shows the graph between focal to aperture width ratio (f/W) and rim angle (in degree), which represent that increase in rim angle means decrease in focal–aperture ratio and also low value of focus–aperture ratio means less spreading of reflected sun rays. A large rim angle means large aperture area which increases the cost of system. In this study, we considered rim angle 70°.

Trough curvature length can be calculated by the equation

$$T = \frac{H}{2}\left\{\sec\frac{\varphi_r}{2}\tan\frac{\varphi_r}{2} + \ln\left[\sec\frac{\varphi_r}{2} + \tan\frac{\varphi_r}{2}\right]\right\}$$

Latus rectum of parabola,

$$H = 4 * f * \tan\frac{\varphi_r}{2}$$

$$D = 2r_r \sin\theta_m,$$

where θ_m is the half acceptance angle and D is receiver diameter which is required to absorb all the solar rays [12].

Here two receiver tubes have been used in parabolic trough system. These receiver tubes have been located at 30° angle from each other so that reflected solar beam radiation can reach perpendicularly to both of the receiver tubes as the rim angle is 70°.

Two receiver tubes provide two times circulation of heat transfer fluid (water) and more exposure of solar radiation which means more rises in heat transfer fluid (HTF) temperature.

3.1 Concentration Ratio (CR)

Concentration ratio can be calculated by aperture area to the receiver area.

$$\mathrm{CR} = \frac{\text{Aperture area}}{\text{Receiver area}}$$

Effective receiver area = $2 * \varphi_r * r * 1$. Here two receiver tubes have been used. φ_r, r, and 1 are the rim angle, radius of absorber tube and length of receiver tube.

3.2 *Intercept Factor (γ)*

Intercept factor means the amount of the reflected beam which is incident on absorbing receiver surface. This parameter is very useful in determination of energy collected. The diameter of a perfect linear imaging concentrator ($D_{\gamma=1}$) having unity intercept factor is given below (Table 1).

$$D_{\gamma=1} = W * \frac{\sin(0.267)}{\sin \varphi_r}$$

$$\gamma = 1 \quad \text{if } D_o = D_{\gamma=1}$$

$$\gamma = \frac{D_o}{D_{\gamma=1}} \quad \text{if } D_o < D_{\gamma=1}.$$

Table 1 Specification of parabolic trough solar concentrator

S. No	Parameter	Value
1	Aperture width, W	1.74 m
2	Length of parabolic trough, l	1.25 m
3	Rim angle, φ_r	70°
4	Focal length of parabolic trough, f	0.61 m
5	Curved length of parabolic trough, T	1.89 m
6	Latus rectum of parabola, H	2.42 m
7	Depth of parabolic trough, d	0.31 m
8	Rim radius, r_r	0.90 m
9	Concentration ratio, CR	60
10	Receiver external diameter, D_o	0.025 m
11	Receiver internal diameter, D_i	0.023 m
12	Glass cover diameter, D_c	0.041 m
13	Collector orientation	Axis in north-south
14	Mode of tracking	East-west
15	Material used for parabolic trough	Stainless steel
16	Material used for receiver	Stainless steel

4 Optical Efficiency

Optical efficiency is the ratio of the energy absorbed by the absorber tube to the solar energy which is incident upon the collector's area. This is given by the following equation [11].

$$\eta_o = \rho\tau\alpha\gamma[(1 - A_f \tan\theta)\cos\theta]$$

where

ρ Reflectivity of mirror
τ Transmissivity of glass cover
α Absorptivity of receiver
γ Intercept factor
η_o Optical efficiency
θ Incident angle
A_f Geometry factor.

4.1 *Geometry factor, A_f*

Geometry factor measures deduction in aperture area because of abnormal incidence of solar beam radiation and loss of sun rays which gets reflected from the reflector beyond the receiver end. Due to irregular operation some of sun rays gets reflected from the collector's end and can't reach to the receiver.

Geometric Factor (A_f) can be calculated as the ratio of shaded area (A_e) due to end effect and aperture area, A.

$$A_e = fW\left|1 + \frac{W^2}{48f^2}\right|.$$

4.2 *Incident Angle, θ*

Angle of incidence is the angle formed between the sun rays hitting a surface to the surface normal.

$$\cos\theta = \left(1 - (\cos\delta)^2 * (\sin\omega)^2\right)^{1/2}$$

where δ is the declination angle and ω is the hour angle.

5 Experimental Setup

The parabolic trough solar collector has been made of stainless steel (SS304). System consist two absorber tube which is kept at 30° away from the focus of parabola. The incoming solar radiation absorbs by the receiver absorber and converts them thermal energy form. Further, this energy is collected and transported by heat transfer fluid (water) which is circulated through absorber tube. The stainless steel absorber tube is enclosed by a concentric glass tube. The space between them is evacuated. Further for minimizing heat losses, vaccum can be applied in the space between glass and metal tube. Performance of parabolic trough system has been investigated for hot water generation as shown in Fig. 3 in a photograph. The test setup consists of a solar collector, receiver tube, water storage tank of 30-L capacity, and control valves for regulating the water flow rate within the circuit.

Based on all the design parameters, a parabolic trough has been designed in AutoCAD Software as shown in Fig. 4.

Fig. 3 Experimental setup

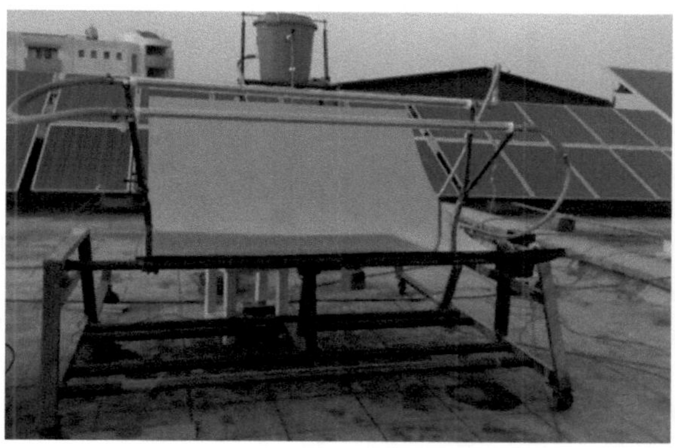

Fig. 4 Parabolic trough concentrator design

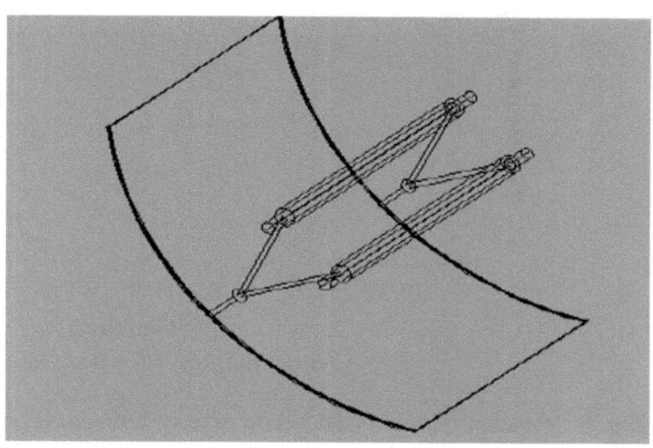

The following results have been obtained with the help of experimental analysis of parabolic trough solar concentrator for hot water generation. The variation of parameters like beam radiations, fluid outlet temperature (temperature of fluid coming out from the receiver tube), and optical efficiency with respect to time are represented graphically. Experimentation was conducted and observations were taken on the system in the AHEC Department at IIT Roorkee. The readings have been taken for an interval of 30 min each. Variation of fluid outlet temperature with respect to time and solar beam intensity on 15 October 2015 has been shown in Fig. 5. The graph shows that fluid temperature increase as time increases. The maximum fluid outlet temperature obtain is 63.1 °C.

Figure 6 shows relationship between fluid outlet temperature with time and solar beam radiation on 7 February 2016. The graph shows that fluid outlet temperature

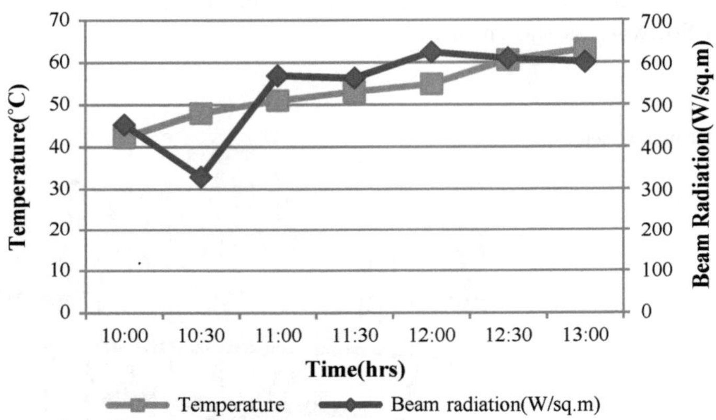

Fig. 5 Variation of fluid outlet temperature with time and beam radiation

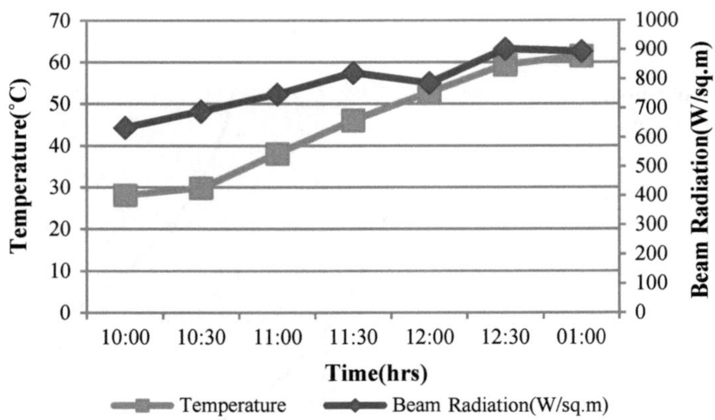

Fig. 6 Variation of fluid outlet temperature with time and beam radiation

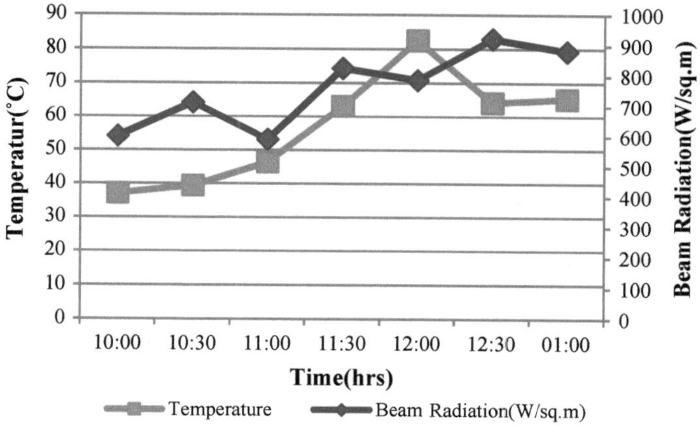

Fig. 7 Variation of fluid outlet temperature with time and beam radiation

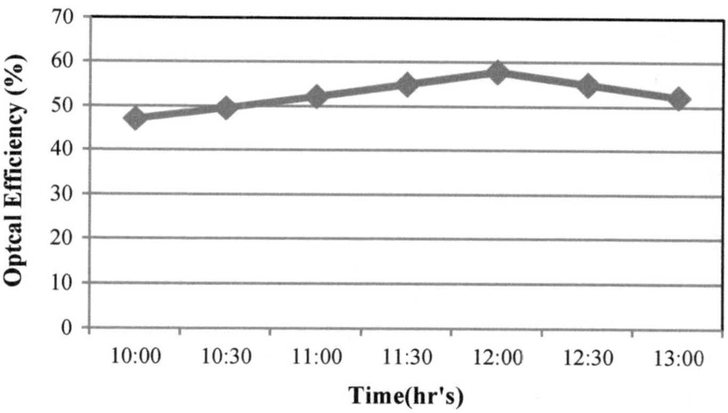

Fig. 8 Variation of optical efficiency with time

increases with time as the solar beam radiation increases. The maximum fluid outlet temperature obtain is 65.3 °C. From the graph, it has been observed that solar beam radiation increase as time increase.

Figure 7 shows variation of fluid outlet temperature with time and solar beam radiation on 8 February 2016. Graph shows mixed data of both clear and cloudy day. The expected maximum solar radiation should occur around noon but it occur around 12:30 pm. Maximum fluid outlet temperature has been found as 82.5 °C.

Figure 8 shows the variation of optical efficiency with time on 15 October 2015 as the time increases the optical efficiency of parabolic trough concentrator increases up to 12o' noon then it decrease as time increases.

Figure 9 shows the variation of optical efficiency with time during 7 February 2016.

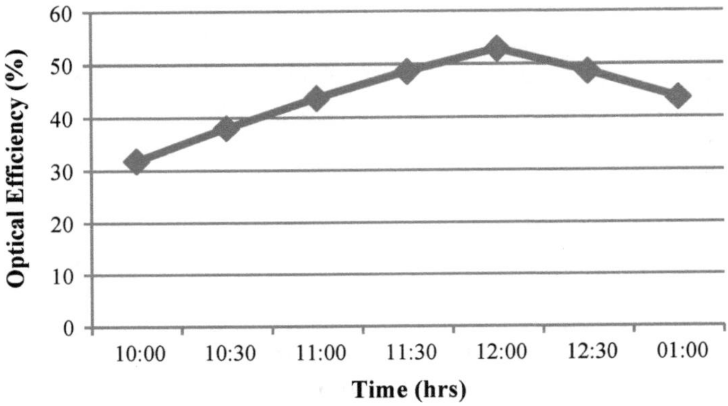

Fig. 9 Variation of optical efficiency with time

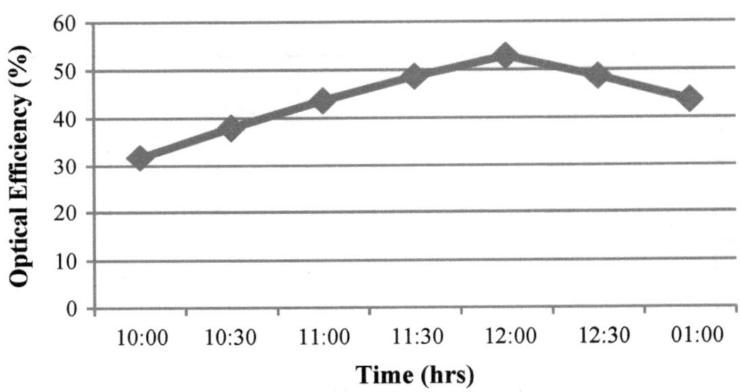

Fig. 10 Variation of optical efficiency with time

Figure 10 shows the variation of optical efficiency with time during 8 February 2016.

For evaluating the performance of parabolic trough concentrator with hot water, graphs have been drawn for determining different parameters such as solar beam radiation, fluid outlet temperature and optical efficiency for several day. The maximum Fluid outlet temperature of 82.5 °C corresponding to solar beam radiation intensity 787 W/m^2 has been obtained with receiver tube made of stainless steel with glass cover. Maximum optical efficiency considering the end effect correction has been obtained 52.83% around noon as during noon incident angle of solar beam radiation is zero.

6 Conclusion

In this study, a parabolic trough solar concentrator has been designed, fabricated and the performance of the designed parabolic trough has been investigated experimentally over a week at Roorkee (India). It has been found that parabolic trough solar collector system is mainly used for water heating, air heat and other applications as well. The maximum value of performance parameter is observed around noon, when the incident sun ray has its maximum value. The value of optical efficiency varies throughout the day and its value was increased up to 52.83% at noon and then decreased. It has been found that maximum fluid outlet temperature is obtained in stainless steel receiver tube as 82.4 °C. The value of optical efficiency depends upon material properties, incident angle of solar beam radiation. It was observed that instantaneous sun ray and velocity of wind is main factor to influence receiver temperature.

References

1. A. Fernandez-Garcia, E. Zarza, L. Valenzuela, M. Prez, Parabolic trough solar collector and their applications. Renew. Sustain. Energy Rev. **14**, 1695–1721 (2010)
2. S.A. Kalogirou, Solar thermal collector and applications. Progress in energy combustion. Science **30**, 231–295 (2004)
3. S.A. Kalogirou, Parabolic trough collector system for low temperature steam generation design and performance characteristics. Appl. Energy **55**, 1–19 (1996)
4. E.A. Mohamed, Designing and testing of solar parabolic concentrating collector. J. Renew. Energy Power Qual. (2013)
5. A. Thomas, V. Guven, Parabolic trough design construction and evaluation. J. Energy Procedia **34**, 401–416 (1993)
6. M.K. Islam, M. Hasanuzzaman, N.A. Rahim, Modelling and analysis the different parameter on a parabolic trough concentrating solar system (2014)
7. Y. Zhang, Q. Zhong-Zhu, L. Peng, G. Wenwer, L. Qiming, H. Jia, *Calculating the Optimum Tilt Angle for Parabolic Solar Trough Concentrator with the North-south Tilt Tracking Mode* (IEEE, 2013)
8. J.M. Valancia, J.R. Avila, O.A. Jasamillo, J.O. Aguilar, Design construction and evaluation of parabolic trough collector as demonstrative prototype. Energy Procedia **57**, 989–998 (2014)
9. M.G. Tayade, R.E. Thombre, S. Dutt, Performance evaluation of parabolic trough. J. Sci. Res. Publ. **5**, 2250–3153 (2015)
10. F. Sulaiman, N. Abdullah, B. Singh, A simulated design and analysis of a solar thermal parabolic trough concentrator. Int. J. Eng. Res. Technol. (2012)
11. A. Kumar, S. Chand, A.O. Umraov, Design and analysis for 1 MWe parabolic trough solar collector plant based on DSG method. Int. J. Eng. Res. Technol. (2013)
12. J. Duffie, W. Beckman, *Solar Engineering of Thermal Process* (Wiley, New York, 2006)

Part III
Receiver and Heat Exchanger

Study on Design of Cavity Receiver of Concentrating Solar Power Plants—A Review

Aditi Garg and R.P. Saini

Abstract The concentrating solar power technology has achieved rapid development in the world. One of the important components of such technology is cavity receiver as it affects the efficiency of the entire power plant. The solar cavity receiver is a photo-thermal conversion component of solar power plants, which heats the working fluid contained in it by absorbing solar radiation. In this paper, an effort has been made to review the work carried out by various researchers on different cavity receiver designs. Based on literature survey, it has been found that cavity geometry has a significant effect on overall flux distribution. The effect of rotation has been observed to be less than 1% for overall receiver efficiency calculations. The thermal losses occurring from the cavity receiver have been found to be affected by various parameters like thermal emissivity, mean fluid temperature, opening ratio and receiver inclination. The wind direction and fluid flow rate also have a considerable influence on the thermal performance of the cavity receiver. It has been realized that there is still scope for developing designs of an efficient cavity receiver which could minimize the losses and hence, enhances the overall efficiency of the plant. The identified gaps in this area for further research have also been discussed in this paper.

Keywords Solar energy · Cavity receiver · Heat loss · Radiation flux

A. Garg (✉) · R.P. Saini
Alternate Hydro Energy Centre, Indian Institute of Technology Roorkee, Roorkee 247667, Uttarakhand, India
e-mail: aditigarg107@gmail.com

R.P. Saini
e-mail: rajsafah@iitr.ac.in

L. Chandra and A. Dixit (eds.), *Concentrated Solar Thermal Energy Technologies*, Springer Proceedings in Energy, https://doi.org/10.1007/978-981-10-4576-9_7

1 Introduction

The concentrating solar power technology plays a crucial role in conversion of solar energy into electricity. It has been used in parabolic trough, Fresnel system, central tower and dish. One of the key components of these solar thermal plants is receiver. Solar radiations are absorbed by the fluid contained in the receiver and then this fluid is further used for power production and heating applications. According to the type of concentrator, there can be a linear focus on receiver or a point focus on receiver. Parabolic trough collector and Fresnel system employs a linear focus on receiver while parabolic dish and central towers employ point focus on receivers. Among the various designs of existing receivers, one of the designs of concern is of cavity receiver. It consists of a large opening on the front face of the receiver which provides an advantage of increased surface area and reduced heat loss. Working fluid can attain higher temperature in cavity receiver. The performance of receiver affects the overall efficiency of the solar thermal power plants. It has been realized that the primary effect of cavity geometry is to vary the radiation flux distribution on the inside walls.

In this paper, a study on the various designs of receiver cavities used in both linear and point concentrators.

2 Cavity Receivers Used in Linear Focusing Concentrators

The linear focusing collectors track the sun along a single axis and focus irradiance on a single receiver. Linear parabolic trough collector and Fresnel system uses a line focusing receiver. The cavity receiver provides a well-insulated enclosure and a small aperture for absorption and reflection of solar energy on the heating surface of cavity. Various works done on these cavity receivers has been summarized in Table 1.

2.1 Cylindrical Cavity Receiver

One of the most common geometry of cavity receiver is a cylinder whose lower part comprising of a cavity receives the reflected solar radiation. Boyd et al. [1] designed a cavity in which the flow of heat transfer fluid (HTF) takes place through the annular channel in the cavity providing an advantage of low flux density and low thermal losses. The cavity design by Barra et al. [2] consisted of copper pipes soldered on the heating surface of the cavity and a removable Pyrex glass at the aperture entrance. Conversion efficiency was observed to be affected by the mean receiver oil temperature, mass flow rate, and inlet temperatures of heat transfer

Table 1 Summary of work on cavity receivers used in linear focus concentrators

Shape of cavity receiver	Work done	Conclusions
Cylindrical cavity receiver	Theoretical performance analysis	1. Low flux density of absorbed solar radiations 2. V-shaped entrance reduces convection and radiative losses
Triangular cavity receiver	Heat transfer analysis	1. High optical efficiency 2. Heterogeneous heat flux distribution 3. Enhanced heat transfer with rectangular fins 4. Outlet temperature of HTF depends on mass flow rate and direct normal irradiance
Trapezoidal cavity receiver	Thermal performance	1. Thermal efficiency decreased with increase in concentration ratio 2. Enhanced thermal performance with selective surface coated absorber 3. The efficiency with the round pipe (multi-tube) absorber was 2–8% higher as compared to rectangular pipe absorber

fluid. Bader et al. [3, 4] initially proposed a cylindrical shaped cavity receiver consisting of a tubular absorber with air as the working fluid which was further improved by incorporating corrugated absorber tube. It was found that increase in the HTF mass flow rate and use of a corrugated absorber tube both reduced the heat loss from the receiver and increased the collector efficiency.

2.2 Triangular Cavity Receiver

Chen et al. [5] designed a triangular cavity receiver and observed that the heat losses from the cavity increased with the increase of angle of inclination and wind speed. The heat lost from the designed cavity absorber has been found to be equivalent to that of the evacuated tube in windless conditions. The cavity design has been further improved by Xiao et al. [6, 26] by addition of glass cover at aperture which helped in reduction of heat losses. The triangular cavity shape facilitated reflections of the sunlight in the cavity. For complete utilization of solar energy, mass flow rate of heat transfer fluid should be changed with direct normal irradiance. Analysis of optical properties of cavity has been carried out by Chen et al. [7] and it was found that the optimal optical properties could be achieved with appropriate aperture width, depth-to-width ratio, and offset distance from focus of triangular cavity absorber.

2.3 *Trapezoidal Cavity Receiver*

Reynolds et al. [8] described experimental techniques for the investigation of the heat losses from the trapezoidal cavity absorber and the flow visualization technique for capturing the flow patterns inside the cavity. Singh et al. [9, 10] studied the effect of receiver shape, concentration ratio and surface coating on the thermal performance of trapezoidal absorber cavity. It was observed that thermal efficiency was enhanced by the surface coatings while it decreased with increase in concentration ratio of the collector. They also found that thermal efficiency increased by 2–8% using a round absorber tube over a rectangular absorber tube. Oliveira et al. [11] analyzed and optimized a trapezoidal cavity receiver for a linear Fresnel solar collector for four parameters, mainly number of absorber tubes, inclination angles of lateral walls of the cavity, cavity depth, and rock wool insulation thickness. The radiation losses were observed to increase with receiver depth as cover and lateral wall surface area increased and heat transfer coefficient decreased with insulation thickness. Mallick et al. [12] studied the effect of Grashof number, absorber angle, aspect ratio, surface emissivities, radiation–conduction number and temperature ratio through the proposed solar trapezoidal cavity model.

The effect of combined Nusselt number on Grashof number and absorber angle was found to be negligibly small. A decrease in Nusselt number has been observed with increase in aspect ratio (ratio of width and depth of cavity) and temperature ratio (ratio of bottom and top surface temperature of cavity), while an increases was observed with surface emissivity. Manikumar et al. [13] analyzed an inverted trapezoidal air cavity having a multi-tube absorber with plate and without plate underneath for various values of gaps between the tubes and depths of the cavity. The thermal efficiency in the case of plane surface was found to be more than that of tube surface. The value of overall heat loss coefficient and convective heat transfer coefficient were observed to increase with gaps between the tubes and the tube temperature.

3 Cavity Receiver Used in Point Focusing Concentrators

Collectors track the sun along two axes and focus irradiance at a single point receiver. This allows good receiver efficiency at a high temperature. The paraboloidal dish and central towers are the point focusing concentrators. Various works done on these cavity receivers has been summarized in Table 2.

Table 2 Summary of work on cavity receivers used in point focus concentrators

Shape of cavity receiver	Conclusions
Cylindrical cavity receiver	1. Cavity-aperture ratio and aspect ratio should be maximized within the limitations of cavity size and operating temperature to enhance the absorption efficiency 2. Convective loss increases with mean receiver temperature and decreases with increase in receiver inclination 3. Convective losses were higher for wind induced conditions than the no-wind conditions for all inclination angles of receiver 4. The temperatures of heat transfer fluid in laminar and turbulent conditions for tightly packed cavity receivers were higher than for medium and loosely packed cavity 5. Effect of rotation was observed to be less than 1% on receiver efficiency
Prism-shaped cavity receiver	1. Non uniform heat flux distribution in the cavity and boiling tubes 2. Change in the wind angle and wind direction directly affected the heat loss from receiver. Maximum heat loss observed when receiver exposed to side-on wind 3. A sharp increase in thermal efficiency was observed with major percentage of losses by convection during the early start-up period 4. As the depth of receiver increased, thermal efficiency increased followed by a decrease. An opposite variation trend has been observed with the heat losses 5. The radiative and convective heat losses decreased as thermal emissivity increased
Hemispherical cavity receiver	1. Natural convection heat loss affected by inclination and shape of the receiver 2. Convection heat losses reduced in modified cavity receiver 3. Maximum heat losses by head-on wind 4. Natural convection heat losses decreased with receiver inclination while opposite trend was observed with forced convection heat losses 5. Radiation heat losses varies with operating temperature and cover emissivity of cavity receiver

3.1 Prism-Shaped Cavity Receiver

Dong et al. [14] presented a combined calculation method for evaluation of the thermal performance of the cavity receiver under different wind environments for a typical shaped cavity. The change in the wind angle wind direction and the speed of air in the region of the boiling tubes affected the heat losses from the receiver. Tu et al. [15] studied the effect of depth of saturated water/steam cavity receiver on thermal performance and found uniformly distributed heat flux and wall temperature of the boiling panels with receiver depth. An increase followed by a decrease has been observed in thermal efficiency as the receiver gets stretched. Wei et al. [16] presented a modified combined method for simulation of the thermal performance of a saturated water/steam solar cavity receiver. Both the radiative and the convective heat losses from the cavity receiver were observed to reduce with increase in thermal emissivity.

Hence, an enhancement in receiver's efficiency was obtained. Fang et al. [17] studied numerically the thermal performance of prism-shaped solar cavity receiver during three kind of start-up process and observed a sharp increase in thermal efficiency and change in the temperature of air during the early start-up period.

3.2 Hemispherical Cavity Receiver

Kumar et al. [18] studied the natural convective heat loss from cavity receiver, semicavity receiver, and modified cavity receiver. It was observed that the orientation and geometry of the receiver strongly affected the natural convection heat loss. The modified cavity receiver was found to have less convection heat losses when compared with the other receivers.

Reddy and Kumar [19] proposed a three-dimensional numerical model for analyzing natural convection heat loss from modified cavity receiver without insulation of fuzzy focal solar dish concentrator. A comparison of two-dimensional and three-dimensional natural convection heat loss from the cavity receiver indicated that latter can be used for an accurate estimation of natural convection heat loss at lower inclination of the receiver while the former model can be used at higher angle of inclination of the receiver.

Veershetty et al. [20] developed a flux distribution model for a fuzzy focal solar dish concentrator system and analyzed the thermal performance of the modified cavity receiver at various wind conditions. The head-on wind was found to significantly influence the heat losses from the receiver. The maximum theoretical efficiency of solar dish collector was reported to be higher at no-wind conditions followed by side-on and head-on wind conditions at wind speed of 5 m/s.

Vikram et al. [21] studied the effect of various parameters on the combined natural convection and surface radiation heat losses from the modified cavity receiver. It was found that the natural convection heat losses were higher at receiver inclination angle, $\beta = 0°$ (receiver facing sideward) and lower at $\beta = 90°$ (receiver facing down) whereas an opposite trend was observed with the forced convection heat loss. The radiation heat losses were found to vary with operating temperature, cavity cover emissivity and marginal variation with inclination angle.

3.3 Cylindrical Cavity Receiver

Wang et al. [14] reported an inverse design method for cylindrical shaped cavity receivers with the cylindrical surface receiving major heat flux. In the peak flux regions, the peak temperatures found were mitigated by the use of a well-designed impinging system without any over pressure drop problem.

Hathaway et al. [22] carried out a numerical simulation based on the Monte Carlo Ray Tracing (MCRT) method to explore the performance of a cylindrical cavity receiver. It was observed that for the best performance in terms of absorption efficiency, the cavity-aperture diameter ratio and aspect ratio should be maximized within the limitations of size and operating temperature. Neber and Lee [23] proposed a receiver design for a Brayton engine-based solar power generation system. It was analyzed that with decrease in aperture diameter, an increase in concentration ratio was needed to produce the required amount of electricity. The increased in concentration ratio enhanced the total efficiency.

Singh [24] developed a simulation for investigating the heat transfer behaviour of a solar cavity receiver. It was observed that working fluid temperatures for both laminar and turbulent conditions were higher in tightly packed cavity receivers than in the medium and loosely packed cavity receivers. The working fluid temperatures and net heat flux has been found to vary with the aperture size of the cavity receiver.

Wu et al. [25] investigated the variation of natural convection losses with rotation of cavity receivers. The effect of rotation on heat transfer was dependent on the receiver orientation. The effect of rotation on the heat losses was observed to be less than 1%. Therefore, for calculating overall efficiency of receiver, it could be neglected.

Xiao et al. [6, 26] designed a tube-cavity solar receiver and conducted experiments for studying the effect of mass flow rate and flow direction on the performance of receiver. The counter current exhibited better performance than that of the parallel flow current and it was observed that the temperature can be controlled by increasing flow rate. The use of glass cover was found to have a positive influence on the air outlet temperature and hence, reduced heat losses.

4 Conclusions

In this paper, an attempt has been made to study the various designs of receiver cavities used in various concentrating solar power technologies. It has been found that inclination and design of the cavity receiver strongly affects the efficiency. The following points have been concluded from various designs studied.

(1) The cavity receivers provide the advantage of low flux density of absorbed solar radiations which reduces temperature gradients and thermal stresses.
(2) The selective surface coated absorber and provision of rectangular fins on the dorsal side of the cavity helped in enhancement of thermal efficiency.
(3) The triangular shaped cavity has highest optical efficiency with heterogeneous heat flux distribution.
(4) Natural convection heat loss from the receiver cavities tend to decrease as the inclination increases from 0 to 90° while an opposite trend was observed with forced convection heat losses. For all receiver inclination angles, convective losses were higher in wind induced conditions than in the no-wind conditions. Radiation heat losses vary with operating temperature and cover emissivity of cavity receiver.

5 Future Scope

(1) The future work involves the study and investigation on change in the structure and material of fins used in the receiver cavities of linear concentrators.
(2) The trapezoidal cavities have not been so far tested with the parabolic trough concentrator. A comparison between the thermal performance of the triangular and trapezoidal cavities has not been conducted so far.
(3) Fewer studies on estimation of optimum aperture size of cavity receiver have been conducted so far.
(4) More study on the effect of wind directions on the geometry of the cavity is required. As in some of literature reviews due to the variations of the opening ratios, head-on wind maximized the heat losses while in other side-on wind enhanced the losses.
(5) Various correlations predicted for the estimation of Nusselt number and heat loss coefficients have applications limited only to certain geometrical and operating conditions. Research on developing such correlations is required.

References

1. O.A. Boyd, R. Gajewski, R. Swift, A cylindrical blackbody solar energy receiver. Sol. Energy **18**, 395–401 (1976)
2. O.A. Barra, L. Franseschi, The parabolic trough plants using black body receivers: experimental and theoretical analyses. Sol. Energy **34**, 163–171 (1982)
3. R. Bader, M. Barbato, A. Pedretti, A. Steinfeld, An air-based cavity receiver for solar trough concentrators. J. Sol. Energy Eng. **132**(2), 031017–031017-7
4. R. Bader, M. Barbato, A. Pedretti, A. Steinfeld, An air-based corrugated cavity-receiver for solar parabolic trough concentrators. Appl. Energy **138**, 337–345 (2015)
5. F. Chen, M. Li, C. Xu, L. Wang, in *Study on Heat Loss Performance of Triangle Cavity Absorber for Parabolic Trough Concentrators* (IEEE, 2013), pp. 683–689
6. G. Xiao, L. Yan, M. Ni, C. Wang, Z. Luo, K. Cen, Experimental study of an air tube-cavity solar receiver. Energy Procedia **61**, 496–499 (2014)
7. F. Chen, M. Li, R.H.E. Hassanien, X. Luo, Y. Hong, Z. Feng, M. Ji, P. Zhang, Study on the optical properties of triangular cavity absorber for parabolic trough solar concentrator. Int. J. Photoenergy **2015**, 1–9 (2015). Article ID 895946
8. D.J. Reynolds, M.J. Jance, M. Behnia, G.L. Morrison, An experimental and computational study of the heat loss characteristics of a trapezoidal cavity absorber. Sol. Energy **76**, 229–234 (2004)
9. P.L. Singh, R.M. Sarviya, J.L. Bhagoria, Thermal performance of linear fresnel reflecting solar concentrator with trapezoidal cavity absorbers. Appl. Energy **87**, 541–50 (2010)
10. L. Singh, R.M. Sarviya, J.L. Bhagoria, Heat loss study of trapezoidal cavity absorbers for linear solar concentrating collector. Energy Convers. Manag. **51**, 329–337
11. A.C. Oliveira, J. Jorge Facao, Numerical simulation of a trapezoidal cavity receiver for a linear Fresnel solar collector concentrator. Renew. Energy **36**, 90–96 (2011)
12. T.K. Mallick, S.K. Natarajan, K.S. Reddy, Heat loss characteristics of trapezoidal cavity receiver for solar linear concentrating system. Appl. Energy **93**, 523–531 (2012)

13. R. Manikumar, R. Palanichamy, A.V. Arasu, Heat Transfer analysis of an elevated linear absorber with trapezoidal cavity in the linear fresnel reflector solar concentrator system. J. Therm. Sci. **24**(1), 90–98 (2015)
14. X.W. Dong, J.B. Fang, J.J. Wei, Y.S. Wang, Thermal performance simulation of a solar cavity receiver under windy conditions. Sol. Energy **85**, 126–138 (2011)
15. N. Tu, J.B. Fang, J.J. Wei, Numerical investigation of start-up performance of a solar cavity receiver. Renew. Energy **53**, 35–42 (2013)
16. J.J. Wei, N. Tu, J.B. Fang, Numerical study on thermal performance of a solar cavity receiver with different depths. Appl. Therm. Eng. **72**, 20–28 (2014)
17. J.B. Fang, J.J. Wei, N. Tu, Numerical investigation on uniformity of heat flux for semi-gray surfaces inside a solar cavity receiver. Sol. Energy **112**, 128–143 (2015)
18. N.S. Kumar, K.S. Reddy, Comparison of receivers for solar dish collector system. Energy Convers. Manag. **49**, 812–819 (2008)
19. N.S. Kumar, K.S. Reddy, An improved model for natural convection heat loss from modified cavity receiver of solar dish concentrator. Sol. Energy **83**, 1884–1892 (2009)
20. G. Veershetty, S.K. Natarajan, K.S. Reddy, Experimental performance investigation of modified cavity receiver with fuzzy focal solar dish concentrator. Renew. Energy **74**, 148–157 (2015)
21. T.S. Vikram, K.S. Reddy, G. Veershetty, Combined heat loss analysis of solar parabolic dish—modified cavity receiver for superheated steam generation. Sol. Energy **121**, 78–93 (2015)
22. B.J. Hathaway, W. Lipinski, J.H. Davidson, Heat transfer in a solar cavity receiver: Design considerations. Numer. Heat Transfer Part A **62**, 445–461 (2012)
23. M. Neber, H. Lee, Design of a high temperature cavity receiver for residential scale concentrated solar power. Energy **47**, 481–487 (2012)
24. Y. Singh, *Thermal Analysis of a Solar Cavity Receiver*. Master of Science, Mechanical Engineering, Lehigh University, United States, pp 1–89 (2012)
25. W. Wu, L. Amsbeck, R. Buck, N. Waibel, P. Langner, R.P. Paal, On the influence of rotation on thermal convection in a rotating cavity for solar receiver applications. Appl. Therm. Eng. **70**, 694–704 (2014)
26. X. Xiao, P. Zhang, D.D. Shao, M. Li, Experimental and numerical heat transfer analysis of a V -cavity absorber for linear parabolic trough solar collector. Energy Convers. Manag. **86**, 49–59 (2014)

Experimental Investigation of Solar Steam Generator Based on Evacuated Tube for Heating and Humidification

Sajith Lawrance, Anunaya Saraswat and Avadhesh Yadav

Abstract In this paper, experimental investigation of solar steam generator based on evacuated tube for heating and humidification has been carried out. The experimental setup consisting of 40 evacuated tubes, a header, and a duct. Water in the header is heated up and converted to steam by using solar energy collected by the evacuated tubes. Steam was generated for 3 h from 13:00 to 16:00 h. The generated steam from the collector is mixed with the ambient air flowing through the duct. The air-flow rate was fixed at 52.31 kg h^{-1}. The conditions at the outlet of the duct are recorded after every half an hour. The maximum temperature difference and humidification rate of 6 °C and 0.580 kg h^{-1} was obtained at 14:00 h, respectively. The results show great potential of evacuated tube collectors for winter air conditioning.

Keywords Solar air conditioning · Solar energy · Evacuated tube collector · Heating and humidification · Winter air conditioning

1 Introduction

The demand for electricity is increasing with rising population, industrialization, and modernization. Developing countries find themselves short in meeting this constantly increasing demand. Most of the power plants still use fossil fuels to produce electricity. These fossil fuels are at the verge of extinction; again burning of these fuels cause environmental problems like air pollution, soil pollution, etc. The above reasons are forcing us to find an alternative energy source, which is renewable and nonpolluting. Solar energy being the most potential and clean energy has attracted scientists through decades.

Evacuated tubes have vacuum around the absorber tube, which reduces convection heat loss thus they better performance than flat plate collectors particularly

S. Lawrance (✉) · A. Saraswat · A. Yadav
NIT Kurukshetra, Kurukshetra, Haryana 136119, India
e-mail: sajithlaw@gmail.com

L. Chandra and A. Dixit (eds.), *Concentrated Solar Thermal Energy Technologies*, Springer Proceedings in Energy, https://doi.org/10.1007/978-981-10-4576-9_8

for high temperature application [1]. After the development of sputter coating machines used for applying selective coating on the absorber tube the market share of evacuated tube collector is rapidly growing.

Following are some of the studies carried out on evacuated tubes till date. A study was carried out to calculate the solar radiation received by cylindrical absorber inside evacuated tube [2]. A study was carried out to corelate the natural flow of fluid inside a single ended evacuated tube collector the result showed that the natural circulation could be corelated with collector inclination, solar intensity, tank temperature, and aspect ratio of the tube [1]. Morrison et al. [3] developed different heat extraction methods for evacuated tubes also numerically studied water circulation behavior in single ended evacuated tube. Evacuated glass tube was investigated numerically and experimentally. The results showed that the performance was related to shape of absorber, incidence angle of radiation and arrangement of tubes [4]. Thermal performance of evacuated tube is evaluated using energy equation also influence of air between copper and absorber tube on efficiency is studied [5]. Studied optimal tilt angle for different types of evacuated tubes and found that tilt angle has to be less than latitude angle [6]. A setup was fabricated for air heating with evacuated tubes; it was also integrated with phase change material [7].

Evacuated tube has been studied since. 1950s [8], still no work has been carried out to produce steam with evacuated tube collector, also no work has been reported on heating and humidification of air with ETC. Heating and humidification is necessary for applications like winter air conditioning, green house, etc. In this paper an experimental investigation of solar stem generator based on evacuated for heating and humidification is presented. The setup is installed at NIT Kurukshetra, Haryana, India (29° 58′ North and 76° 53′ East).

2 System Description

Solar steam generator based on evacuated tube for heating and humidification is shown in Fig. 1. The system consists of 40 evacuated tubes attached to a header. The high head of the header opens to the duct. A blower is attached to the duct as shown in figure. Evacuated tubes and header are inclined at 5° and 15° from the horizontal, respectively. The header and the forty evacuated tubes are filled with 80 L of water. Two photographic view of the setup is shown in Figs. 2 and 3. The main components of the system are as follows.

2.1 Evacuated Tubes

Evacuated tubes are made up of two coaxial borosilicate glass tubes fused at one end. The dimensions of the tubes are inner diameter 37 mm, outer diameter 47 mm

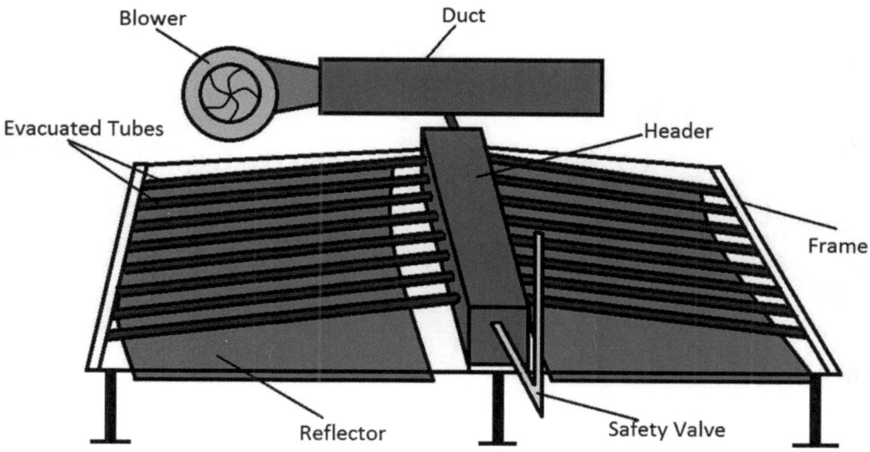

Fig. 1 Schematic diagram of the setup

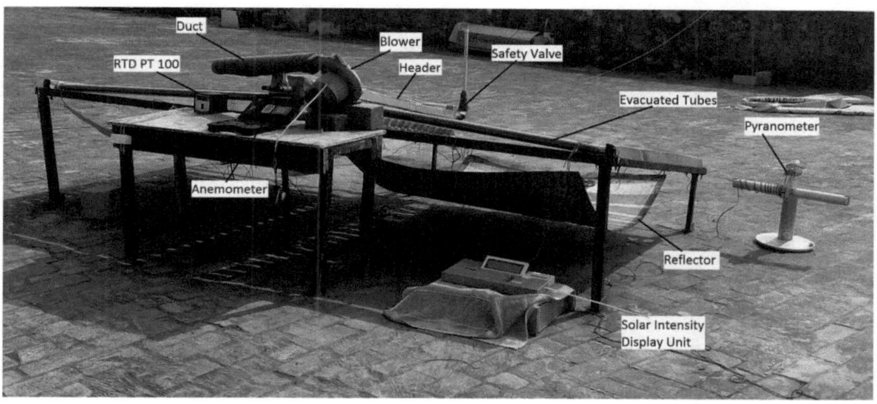

Fig. 2 Photograph of the setup (view 1)

and length 1500 mm each. There is vacuum between the two layers of the glass to reduce the heat transfer loss. The outer tube is transparent and has high transmissivity. The outer surface of the inner tube is coated with a layer of aluminum nitride (Al-N/Al) this coating has high absorptivity and less emissivity.

2.2 Header

Schematic diagram of the header is shown in Fig. 4. The header is fabricated from a steel box of dimension 120 mm × 120 mm × 1500 mm. This header is insulated

Fig. 3 Photograph of the setup (view 2)

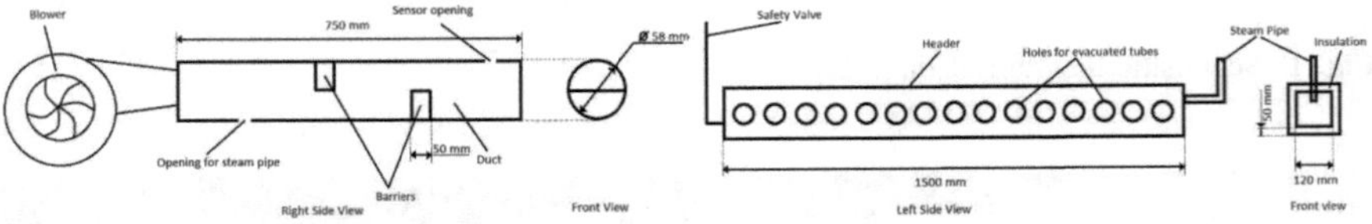

Fig. 4 Schematic diagram of the duct (*left*) and the header (*right*)

with polystyrene from all sides to decrease the heat loss. The thickness of the insulation is 50 mm. A safety valve is attached to the low head of the header. A pipe was provided at the high head of the header for steam outlet.

2.3 Duct

A newly designed duct is fabricated for this setup. The schematic diagram of the duct is shown in Fig. 4. It is made of polyvinyl chloride material. The duct has length of 750 mm and diameter of 58 mm. Blocks of polystyrene material is fixed inside the duct as shown in figure to ensure proper mixing of steam with the flowing air.

2.4 Reflector

There are gap between the evacuated tubes. Some part of the incident solar radiation get lost as these rays do not fall on the tubes and escape through the gap. In order to minimize this loss a reflector is attached under the tubes, which reflect the solar radiations back to the tubes. The reflector has dimension of 1550 mm × 1200 mm.

This reflector is made from galvanized aluminum sheet having zinc coating on the surface. It has reflectivity of 86%. The zinc coating helps in increasing the reflectivity of the sheet.

3 Measuring Devices and Instruments

In this experiment the data recorded were temperature of water at low head and high head of header, ambient dry bulb and wet bulb temperatures, duct outlet dry bulb and wet bulb temperatures and solar radiation. The air-flow rate was kept constant. The details of devices used for measurements are shown in Table 1.

4 System Working

The setup was exposed to the sun one day before the readings were taken. Natural circulation of water taking place in header and evacuated tubes are shown in Fig. 5. Water in the evacuated tubes is heated up by the energy received from the sun. Due to this heating and inclination of tubes, thermosyphon effect takes place in the tubes, thus hot water from pipes moves to the header and cold water from header rushes into the tubes. The inclination of the header with the horizontal plane forces

Table 1 Specification of measuring instruments

Measurement	Devices	Range	Accuracy
Solar intensity	Pyranometer	0–1400 W m^{-2}	±2 W m^{-2}
Temperature (dry bulb and wet bulb)	RTD PT 100 temperature sensor	0–200 °C	±0.3 °C
Air velocity	Anemometer	0–45 m s^{-1}	±2%

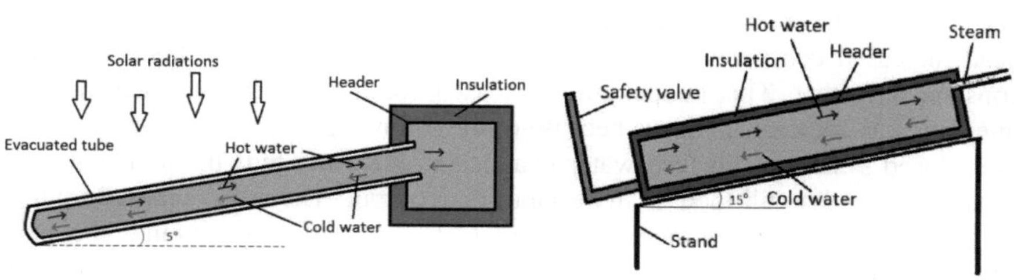

Fig. 5 Thermosyphon in evacuated tubes (*left*) and thermosyphon in the header (*right*)

hot and less dense water to the top of the header. The hot water at the high head of the header gets converted into steam. This steam is then carried into the duct by the steam pipe. The steam gets mixed with the air flowing in the duct thus heating and humidifying the air.

5 Performance Indexes

The performance of the system is measured in terms of its COP (coefficient of performance), which is the ratio of the energy gained by the air to the solar energy received for that duration dt (in seconds). Where m_a is mass of air, h_1 is enthalpy of ambient air, h_2 is enthalpy of air at outlet of duct, I is solar intensity, and A denotes collector area. An example calculation follows. The input and output of the system is calculated for every half an hour and added up to calculate the overall COP of the system for that day, for, e.g., the enthalpy of air at inlet and outlet of the duct at 13:00 h were 44.937 and 56.457 kJ kg^{-1}, respectively, and at 13:30 h were 46.392 and 72.33 kJ kg^{-1}, respectively. The average enthalpy difference came out to be 18.729 kJ kg^{-1} for this half hour period. The air-flow rate was constant at 52.31 kg h^{-1}. The average solar intensity for this period was 720.5 W m^{-2} and the area of collector was 4.44 m^2

$$\mathrm{COP} = (m_a 1000(h_2 - h_1)) \div (IA\mathrm{d}t). \tag{1}$$

6 Results and Discussion

The experiment was carried out at NIT Kurukshetra, India on 6th of February 2016. There was fog till 10:00 in the morning after which the sky was clear. The data were recorded from 10:30 to 16:00. Ambient temperature varied from 8 to 25 °C. The air velocity was fixed at 5.5 m s^{-1}. The steam started generating at 13:00 and was generated for 3 h till 16:00.

Variation of temperature of water at low head (L.H.) and high head (H.H.) of the header with solar intensity and time is shown in Fig. 6. The temperature at the high head increased continuously with time till it reached 101 °C, it almost remained constant after that. This temperature increased continuously even when the solar intensity was decreasing; this is because of the following reasons, the system works as a closed system as no feed water is added to the setup thus the temperature of water will increase till there is more input than output. The setup generates output only after reaching temperature above 100 °C, hence, there is no output at this point and solar energy is delivering input, which is decreasing but still there is input. Thus increasing the high head water temperature. But due to decrease in solar intensity the rate of increase in temperature of water decreased. After the water

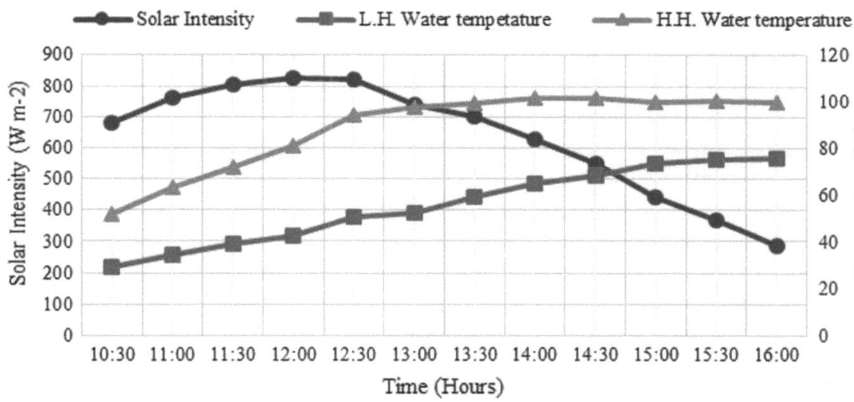

Fig. 6 Variation of water temperature and solar intensity with time

attained temperature of 101 °C the setup started delivering output and thus the temperature of water remained constant. The low head showed almost same curve like the high head the exception was it gained heat by conduction from high head when the solar intensity was diminishing. It reached a maximum temperature of 75.7 °C. The solar intensity reached a maximum value of 827 W m^{-2} at 12:00 h.

The condition of air at outlet of duct and ambient condition of air in terms of its humidity and temperature is shown in Fig. 7. The ambient air temperature (dry bulb temperature) remained between 23.4 and 24.6 °C during the working hours (working hours of setup). The setup was able to maintain the outlet temperature between 30 and 27.4 °C. The ambient air humidity stayed between 8.40 and 9.50 g kg^{-1} during working hours. The setup was able to maintain the outlet humidity between 9.50 and 19.6 g kg^{-1}. The curve of outlet humidity ratio was related to high head water temperature and solar intensity. Initially with rise in temperature of high head water the humidification rate increased but latter on with the considerable drop in solar intensity the humidity at outlet also dropped.

Figure 8 shows the temperature (dry bulb temperature) difference obtained for air at the outlet of duct to ambient air, and the humidification rate achieved by the setup. The temperature difference obtained can be associated with high head temp and solar intensity. Initially when the high head temperature was increasing the temperature difference kept on increasing but with the considerable fall in solar intensity the temp difference started decreasing. At 15:30 the temperature suddenly raised instead of decreasing this was because of sudden fall in ambient temperature of air, which created large difference in temperature of setup water and incoming air, increasing the heat transfer coefficient. The maximum temperature difference obtained was 6 °C at 14:00. The humidification rate obtained can be associated with high head temp and solar intensity. Initially when the high head temperature was increasing humidification rate kept on increasing but with the considerable fall in solar intensity the humidification rate started decreasing. The maximum humidification rate of 0.580 kg h^{-1} was achieved at 14:00 h.

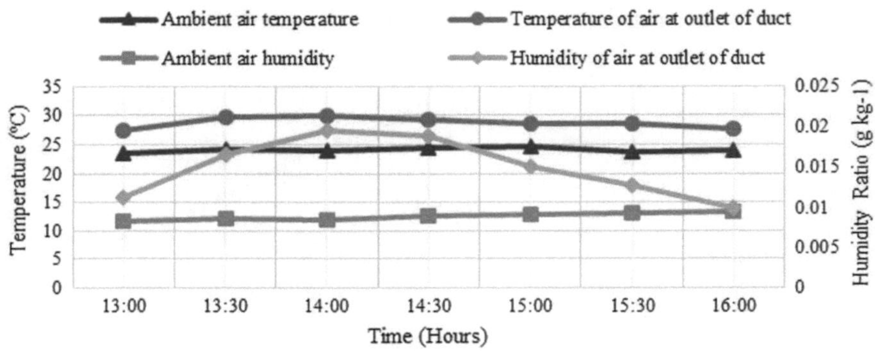

Fig. 7 Variation of temperature and humidity ratio with time at inlet and outlet of the duct

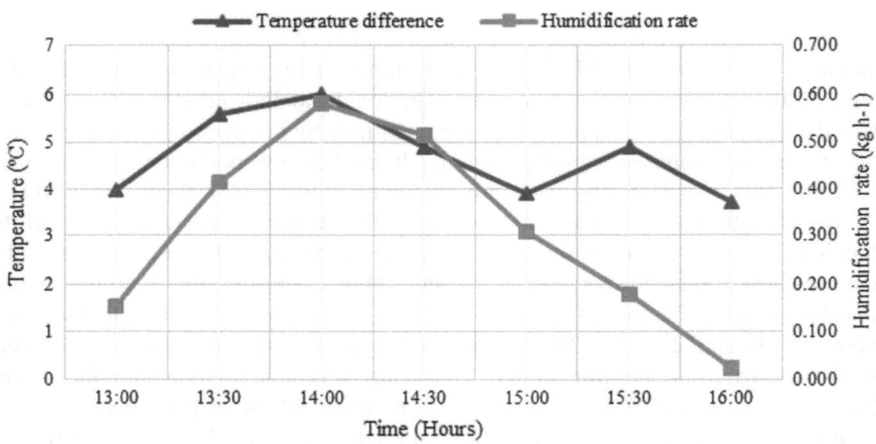

Fig. 8 Variation of temperature difference, humidification rate with time

The COP of the system is calculated using Eq. (1) shown in performance indexes. The overall COP obtained by the setup was 0.0599. Though the COP seems little on the lower side there is another output, i.e., hot water that is not considered in the calculation of COP. Thus the overall efficiency of the system can be improved if we can make use of this hot water.

7 Conclusions and Recommendations

On a typical north Indian winter conditions where ambient temperature varied from 8 to 25 °C the setup was able to maintain a temperature difference of 3.5–6 °C for 3 h. The maximum humidification rate obtained was 0.580 kg h^{-1}. The setup obtained overall COP of 0.0599. The above results show that the steam generator based on evacuated tube has great potential for replacing winter air conditioners but further research need to be done in this field. Some of the future work that can be done is integrating this system with phase change material or other thermal storage system, automatic controlling of the steam and controlling of air-flow rate with sensors.

References

1. I. Budihardjo et al., Natural circulation flow through water-in-glass evacuated tube solar collectors. Sol. Energy **81**, 1460–1472 (2007)
2. R. Perez et. al., Calculating solar radiation received by tubular solar-energy collectors. Sol. Eng. **I**, 699–704
3. G.L. Morrison et al., Water-in-glass evacuated tube solar water heaters. Sol. Energy **76**, 135–140 (2004)
4. Y. Kim, T. Seo, Thermal performances comparisons of the glass evacuated tube solar collectors with shapes of absorber tube. Renew. Energy **32**, 772–795 (2007)
5. L. Ma, Thermal performance analysis of the glass evacuated tube solar collector with U-tube. Build. Environ. **45**, 1959–1967 (2010)
6. R. Tang et al., Optimal tilt-angles of all-glass evacuated tube solar collectors. Energy **34**, 1387–1395 (2009)
7. N. Mehla, A. Yadav, Experimental analysis of thermal performance of evacuated tube solar air collector with phase change material for sunshine and off-sunshine hours. Int. J. Ambient Energy (2015). https://dx.doi.org/10.1080/01430750.2015.1074612
8. M.J. Lighthill, Theoretical consideration on free convection in tubes. Quart. J. Mech. Appl. Math. **6**(4), 398–439 (1952)

Selective Design of an Experiment for Evaluating Air–Water Hybrid Steam Condenser for Concentrated Solar Power

Sumer Dirbude, Nashith Khalifa and Laltu Chandra

Abstract In thermal power plants water-cooled steam condenser is used to reject heat from steam at the turbine outlet. In this system, makeup water of about 1500–3000 L per MWh of electricity generation is required. Therefore, in the arid places, air-cooled condensers are recommended, especially, for concentrated solar thermal power (CSP) plants. It is reported that 10 °C rise in air temperature reduces electricity output by about 4.2% Bustamante et al. (Appl Therm Eng xxx:1–10, 2015) [1]. For steam condensation with air at a temperature of 37 °C, an exergetic efficiency of the condenser is 26% in comparison to 63% for water-cooled steam condenser Blanco-Marigorta et al. (Energy, 36:1966–1972, 2011) [2], Bustamante et al. (Appl Therm Eng xxx:1–10, 2015) [1]. Therefore, air-cooled condenser needs higher initial temperature difference (ITD) between condensing steam and air to achieve a high power output. In dry and hot places, like Rajasthan in India, during summer, air temperature reaches up to 45–50 °C. Whereas, condensing steam at the exit of turbine is available at a pressure of $\sim$0.1 atm and at a temperature of $\sim$45 °C. In view of this, the paper presents design of an experiment for evaluation of an air–water hybrid steam condenser. The proposed concept is based on: (a) the temperature difference between air and condensing steam in dry and arid regions and (b) achieving a high value of Nusselt/Colburn-factor to friction-factor ratio. A plate fin-and-circular tube with staggered arrangement air-cooled condenser designs is selected, as the starting point. Two possible modifications are suggested: (a) Earth (underground) for pre-cooling of ambient air or water and (b) use of water spray for additional cooling of the pre-cooled air. The obtained cold air is employed for heat transfer in condenser. In this paper, the selection and evaluation of characteristic

S. Dirbude (✉) · N. Khalifa
Centre for Solar Energy Technology, Indian Institute of Technology Jodhpur, Jodhpur, Rajasthan, India
e-mail: sumery7@gmail.com; sbdirbude@iitj.ac.in; sumer@iitk.ac.in

L. Chandra
Department of Mechanical Engineering and Center for Solar Energy, Indian Institute of Technology Jodhpur, Jodhpur 342011, Rajasthan, India
e-mail: Chandra@iitj.ac.in

L. Chandra and A. Dixit (eds.), *Concentrated Solar Thermal Energy Technologies*, Springer Proceedings in Energy, https://doi.org/10.1007/978-981-10-4576-9_9

design parameters are presented. Finally, the designed experimental setup using all these aspects is described.

Keywords Experimental design · Concentrated solar power · Air–water hybrid condenser · Heat transfer · Pressure drop

Nomenclature

D	Characteristic dimension
f	Friction factor $= \Delta p / \frac{1}{2}\rho U_{max}^2 (L/D)$
h	Heat transfer coefficient
j	Colburn heat transfer coefficient = $Nu/Re\,\mathrm{Pr}^{1/3}$
k	Thermal conductivity of fluid
N	Number of tube rows
Nu	Nusselt Number $= hD/k$
P_f/D	Fin pitch to tube collar diameter ratio
P_l/D	Longitudinal tube pitch to tube collar diameter ratio
Pr	Prandtl number
P_t/D	Transverse tube pitch to tube collar diameter ratio
Re_D	Reynolds number (tube collar diameter D) $= \rho UD/\mu$
RH	Relative humidity (%)
α	Half contraction cone angle
β	Half diffuser angle
θ	Full-spray cone angle
ρ_m	Water vapor + dry-air mixture density
δ_f	Fin thickness
Δ_p	Pressure drop
Δ	Triangular cross-section
$\sim$	Sinusoidal cross-section

1 Introduction

In thermal power plants, for example, based on solar and nuclear energy, steam condenser is used to reject heat to ambient via a working fluid. Generally, water is used as heat rejecting fluid for condensation of steam. For an efficient operation of steam condenser, the basic parameters are: (a) the temperature difference between air and condensing steam in dry and arid regions and (b) high values of Nusselt/Colburn-factor and friction-factor ratio. Water-cooled condensers are widely used in view of heat capacity and thermal conductivity of fluid. However, the cooling water requirement for condensation of steam, for example, in thermal power plants is 1500–3000 L/MWh of electricity generation [3]. Therefore, in the arid places, air-cooled condensers may be recommended, especially for CSP plants.

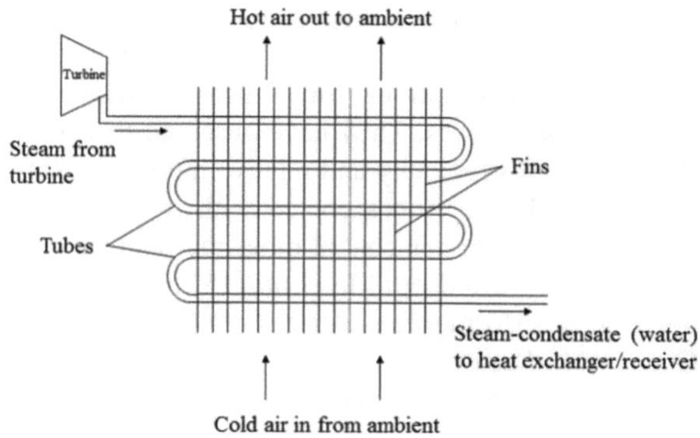

Fig. 1 Schematic of typical air-cooled condenser used in solar thermal power plant

A schematic of typical air-cooled condenser is shown in Fig. 1. In air-cooled condensers, increase in temperature of incoming air reduces the steam turbine output [1]. To increase initial temperature difference (ITD) between condensing steam and ambient incoming air, various methods viz. spray, Munters, hybrid, and deludge are in use (see for, e.g. Kutscher and Costenaro [4]). Kutscher and Costenaro [4] stated that, although deluge method was found effective, corrosion of finned tubes is a disadvantage. On the other hand, spray method was found to give higher electricity production per month among the others. Recently, for instance, Alkhedhair et al. [5] has reported that the addition of relatively small quantity of water in the form of spray into ambient air cools it by 8–12 °C. They suggested to use air velocity (approx. <1 m/s) and droplet size distribution approximately <70 μm as against Tissot et al. [6] who recommend the value of <25 μm. These studies, however, only discuss thermal enhancement of air using water spray and indicate requirements of the spray quality. Earth-based cooling of air is used in air-conditioning (see for instance, El-Agouz and Kabeel [7]). In these, various approaches are being used in the design of earth air heat exchanger (EAHE) for cooling of air. Chandra et al. [8] has established one such approach to design EAHE. Although many studies are done on EAHE, to best of our knowledge, none have found explored this concept for air-cooled condenser applications.

Fin-and-tube condenser (heat exchanger) is widely used under air-cooled category [9, 10]. The heat transfer to air enhancement with minimum pressure drop, with the given ITD, is solely dependent on the geometrical and flow parameters of the heat exchanger. For dry air, Webb and Gray [11], Kim et al. [10], Wang et al. [12], Xie et al. [13], and Wang et al. [14] studies report effect of geometrical and flow parameters in terms of correlations for Nusselt number/Colburn heat transfer coefficient and friction factor. These studies, within their test range, show that the geometrical and flow parameters have significant whereas others have negligible impact on dry-air side heat transfer and pressure drop characteristic. For wet air, on the other hand, most of the studies are devoted to heat transfer and friction-factor correlations for evaporator applications that are found in air-conditioning. For evaporators, tube surface conditions may exist as dry or wet (this condition will exist if tube surface is at temperature lower than dew point temperature of the inlet air) depending on its temperature. For instance, Wang and Chang [9] gave the wet-air side correlations for a heat exchanger with plain fin and staggered arrangement of tubes. They have observed that for wet surface conditions the inlet flow conditions [velocity and relative humidity (RH)] of air increases pressure drop for heat exchanger tubes coated with hydrophilic material when compared with uncoated tubes. For uncoated tubes the inlet flow conditions are found to have no effect on pressure drop. Ma et al. [15] have tested performance of wavy, slit, and louver fins performance under wet conditions. They reported, contrary to Wang and Chang [9], that the RH of inlet air has negligible effect on Colburn *j*-factor and friction factor. Yun et al. [16] have performed experiments on compact finned tube heat exchangers under wet conditions. They report that the *j*-factor is ~33–55% larger for small tube diameters than that of higher tube diameters.

From the literature, it is observed that a combination of earth and water spray-based cooling of air where water can be cooled using earth-based cooling before used as spray, to our knowledge, is not yet studied for the air-cooled condensers. This strategy is proposed to work especially in dry and hot places, like Rajasthan in India, where air temperature reaches up to 45–50 °C and steam at the turbine exit is available at a lower/comparable temperature of ~45 °C. Next, it is observed that there exist many correlations for dry-air side heat transfer and pressure drop. However, very few wet-air side correlations are available for the compact heat exchangers. Also, for wet air, due to limited test results, it is hard to find general conclusion on heat transfer and pressure drop characteristics. A long-term objective of this work, therefore, is to study the effect of tube diameter, fin spacing, tube spacing, RH on heat transfer and pressure drop characteristics for the design of air-cooled steam condenser for CSP applications. In this work, however, a conservative approach is presented for an experiment design for an air–water hybrid steam condenser from the existing dry-air side heat transfer and pressure drop correlations.

2 Experimental Setup: Design Methodology

In this work an experiment design of an air–water hybrid (air + water spray cooled) heat exchanger (AWHX) is presented. It requires specification of various parameters, including transverse tube spacing, longitudinal tube spacing, tube diameter, number of tube rows, fin spacing, fin thickness, fin type (plain or enhanced), etc. The schematic of heat exchanger is shown in Fig. 2. For the design of AWHX the adopted approach is: (a) circular tubes with staggered arrangement is assumed, (b) the condensing steam is indicated as heat flux to AWHX, (c) selection of parameters for AWHX on the basis of figure of merits and finally, (d) design of wind tunnel and test section for experimental evaluation. In the following sub-sections, selection of the parameters of the wind-tunnel and test-section design is given.

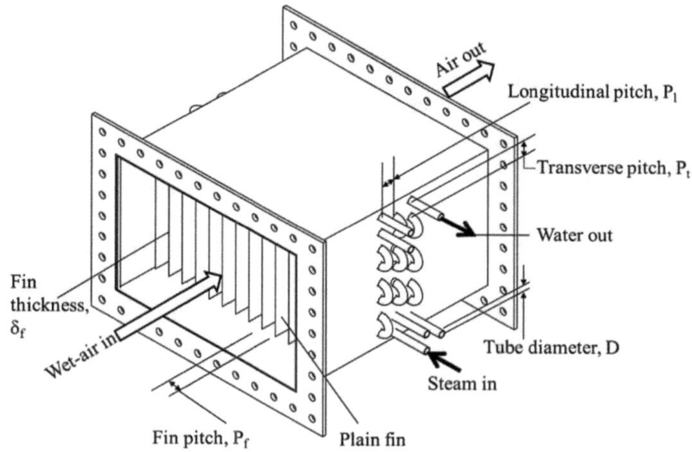

Fig. 2 Schematic of heat exchanger

2.1 Selection of Parameters of AWHX

Selection of AWHX parameters for experimental evaluation requires a rigorous process for data reduction. For the optimal performance of AWHX, heat transfer to pressure drop ratio should be maximum. To select various parameters, as the starting point, various correlations available for Nusselt/Colburn-factor and friction factor are compared. For this purpose:

(a) The correlations of Webb and Gray [11] (Eqs. 1, 6a and 6b), Kim et al. [10] (Eqs. 2, 6c and 6d), Wang et al. [12] (Eqs. 3 and 7), Xie et al. [12] (Eqs. 4 and 8), and Wang et al. [13] (Eqs. 5 and 9), which are given for plain fin and staggered tube arrangements with variable parameters, are selected. The range of operating conditions for these correlations and their uncertainties for various parameters are mentioned in Table 1. From these equations, optimal values of fin pitch, tube pitch, tube diameter are derived to maximize heat transfer to friction ratio within the considered range. Here, the averaged heat transfer (Nu or j) to pressure drop (f) ratio are compared for various Reynolds number (Re) for three number of tube rows obtained from the lowest and highest values of fin pitch, tube pitch, tube diameter mentioned in the range of validity for each correlation. From this, a conservative correlation is selected.

(b) Effect of the number of tube rows on j/f at various Re is compared for the selected correlation. From this, the numbers of tube rows are finalized.

(c) A selection of fin type, for fixed number of tube rows, is subsequently carried outfor AWHX. The experimental [17] and numerical [18] data are used. The ratio of Nu/f at various Re is compared for plain, wavy (Δ: triangular; ∼: sinusoidal), slit fin types at three fin pitches. The fin type is selected as a

Table 1 List of parameters and ranges for staggered arrangement of tubes

Components/parameters	Webb and Gray [11], Kim et al. [10], Wang et al. [12], Xie et al. [13], Wang et al. [14]	Kang et al. [17]	Wang and Chang [9]
Fin type and tube arrangement	Plain and staggered	Plain, wavy (Δ and ~), slit, and staggered	Plain, and staggered
RH	–	–	46–85%
Re	300–24,707	350–15,600	500–5000
N	1–8	3	1–3
D (mm)	6.35–20	10.15	5.50–10.23
P_f/D	0.08–0.641	–	
P_t/D	1.97–2.881	–	
P_t/P_l	0.857–1.654	–	
P_l/D	1.7–2.58	–	
P_f (mm)	1.19–8.7	1, 2.6, 3.2	1.2–1.8
P_t (mm)	17.7–36	25	20–25.4
P_l (mm)	12.4–46	21.65	10–19.05
δ_f (mm)	0.13–0.2		0.1–0.26
Uncertainty (%)			
Nusselt number/Colburn heat transfer coefficient	3.7–10.7	6.4	7.3
Friction factor	6.5–12.6	7.7	7.8

compromise between higher values of *Nu*/*f*, ease of manufacturing and cost. For the selected fin type, experimental values of fin pitches are recommended.

(d) For fixed values of tube pitches, optimum fin thickness is selected.

$$j = \left\{0.14Re_D^{-0.328}(P_t/P_l)^{-0.502}(P_f/D)^{0.0312}\right\}0.991 \left[2.24Re_D^{-0.092}(N/4)^{-0.031}\right]^{0.607(4-N)} \tag{1}$$

$$j = \left\{0.163Re_D^{-0.369}(P_f/D)^{0.0138}(P_t/D)^{0.13}(P_t/P_l)^{0.106}\right\} \times 1.043 \times \left[Re_D^{-0.14}(P_f/D)^{-0.123}(P_t/D)^{1.17}(P_t/P_l)^{-0.564}\right]^{(3-N)} \tag{2}$$

$$j = 0.086Re_D^{P3}N^{P4}(P_f/D)^{P5}(P_f/D_h)^{P6}(P_f/P_t)^{-0.93} \tag{3}$$

where,

$$P3 = -0.361 - [0.042N/\ln(Re_D)] + 0.158\left(N(P_f/D)^{0.41}\right),$$
$$P4 = -1.224 - \left[0.076(P_l/D_h)^{1.42}/\ln(Re_D)\right],$$
$$P5 = -0.083 - [0.08N/\ln(Re_D)],$$
$$P6 = -5.735 + 1.21\ln(Re_D/N),$$

where, D_h is the hydraulic diameter $= \frac{4A_{min}L}{A_o}$ with $A_o = A_f + A_t$ and L is the depth of heat exchanger in air flow direction.

$$Nu = 1.565Re_D^{0.3414}(N \times (P_f/D))^{-0.165}(P_t/P_l)^{0.0558} \quad (4)$$

$$j = 0.394Re_D^{-0.392}(\delta_f/D)^{-0.0449}N^{-0.0897}(P_f/D)^{-0.212} \quad (5)$$

$$f = f_f(A_f/A_o) + f_t[1 - (A_f/A_o)][1 - (\delta_f/P_f)] \quad (6)$$

where,

$$f_f = 0.508\mathrm{Re}_D^{-0.521}(P_f/D)^{1.318}, \quad (6a)$$

$$f_t = \frac{4}{\pi}\left\{0.25 + \left(0.118/[(P_t/D) - 1]^{1.08}\right)Re_D^{-0.16}\right\}[(P_t/D) - 1], \quad (6b)$$

$$f_f = 1.455Re_D^{-0.656}(P_f/D)^{-0.134}(P_t/D)^{1.23}(P_t/P_l)^{-0.347}, \quad (6c)$$

$$f_t = \frac{4}{\pi}\left\{0.25 + \left(0.118/[(P_t/D) - 1]^{1.08}\right)Re_D^{-0.16}\right\}[(P_t/D) - 1], \quad (6d)$$

$$f = 0.0267Re_D^{I1}(P_t/P_l)^{I2}(P_f/D)^{I3}, \quad (7)$$

where,

$$I1 = -0.764 + 0.739(P_t/P_l) + 0.177(P_f/D) - (0.00758/N),$$
$$I2 = -15.689 + (64.021/\ln(Re_D)),$$
$$I3 = 1.696 - (15.695/\ln(Re_D)), \quad (8)$$
$$f = 20.713Re_D^{-0.3489}(N \times [P_f/D])^{-0.1676}(P_t/P_l)^{0.6265},$$

$$f = 1.039Re_D^{-0.418}(\delta_f/D)^{-0.104}(N)^{-0.0935}(P_f/D)^{-0.197}. \quad (9)$$

2.2 Design of Wind Tunnel and Test Section

The proposed experimental setup is shown in Fig. 3. The dimensions of the wind tunnel (for maximum wet-air velocity, 10 m/s with corresponding $Re \approx 6000$ based on heat exchanger outside tube diameter) are obtained using a method mentioned in González Hernández et al. [19] and Brusca et al. [20]. The dimensions of settling chamber, contraction cone ($\alpha = 12°$) and diffuser ($\beta = 3.5°$) are based on the dimensions of cross-section of the test section. Here, the width and height are 0.3 and 0.2 m, respectively. The fan power is calculated for wet-air (mixture density at 85% RH, $\rho_m = 1.25$ m^3/kg) with velocity of 10 m/s from the total pressure drop and volume flow rate in the system. The individual pressure loss coefficients are obtained using equations mentioned in Barlow et al. [21].

Relative locations of various sub-systems, for example, the spray nozzle and the heat exchanger, inside the test section are shown in Fig. 4. A spray nozzle and heat exchanger is placed in a test section having width-to-height ratio of 1.5 by referring to the state-of-the art mentioned, for example, in Biswas et al. [22], Tissot et al. [6] and Chen and Lai [23]. The distance of spray nozzle at 0.3 m from the exit of contraction cone, ensuring uniform flow, is decided from the wind-tunnel data used by Biswas et al. [22]. The grid location, to measure temperature field of moist air, is placed at 0.9 m from the spray nozzle. This distance is larger than spray length [$l = (H/2)/\tan(\theta/2)$ where, H is the test-section height and θ is the full cone spray angle] of 0.56 m for a nozzle with 45° (minimum) full-spray angle. A heat exchanger is located in the test-section comprising a length of 1.5 m [6, 22, 23] for

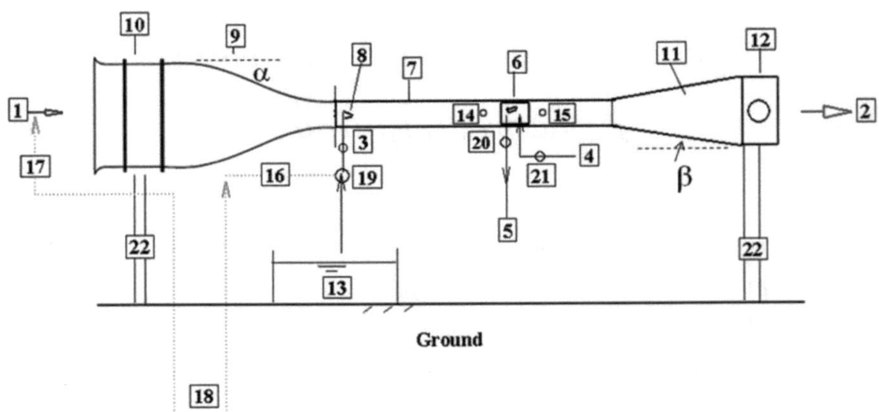

Fig. 3 Proposed experimental setup. *1* Air inlet, *2* air outlet, *3* Pressure sensor, *4* hot water in, *5* cold water out, *6* heat exchanger, *7* test section, *8* water spray nozzle, *9* contraction cone, *10* settling chamber, *11* diffuser, *12* blower/fan, *13* water container for spray, *14* and *15* measurement locations, *16* earth/ground cooled water (not used here) inlet, *17* earth/ground cooled air inlet (not used here), *18* geothermal cooling, *19* water pump, *20* and *21* measurement locations, *22* supports

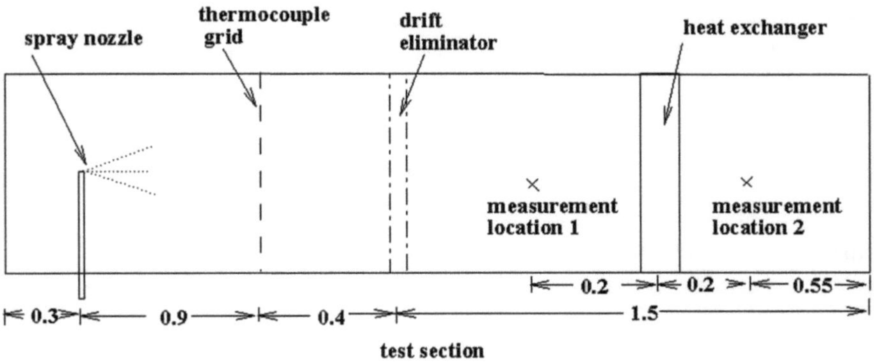

Fig. 4 Details of test section (in meters)

the same width-to-height ratio of 1.5 of the test section. It is placed at 0.75 m from the exit of drift eliminator to ensure uniform flow [22]. The measurement locations, before and after heat exchanger, is kept in the range 0.1–0.3 m [22].

3 Results and Discussions on Experimental Design of Heat Exchanger

As the starting point selection of five AWHX parameters, viz. fin pitch, tube pitches, fin thickness, tube diameter, for experimentation in the range $500 < Re < 6000$ (based on the heat exchanger tube outer diameter) is planned. To study the effect of each parameter at single *Re* may need 25 experiments. Therefore, a rigorous process for data reduction may help in reducing number of experiments to be carried out. In the following paragraphs, results of the procedure mentioned in section—Experimental setup: Design methodology—are given and discussed.

To decide the numerical values of heat exchanger parameters, the optimal values of j_{max} or Nu_{max} and f_{min} are calculated from Eqs. (1)–(5), and Eqs. (6)–(9) (see Table 1), respectively. The obtained values of j_{max} or Nu_{max} and f_{min} are given in Table 2. It is observed that, for the obtained values of j_{max} and f_{min}, the geometrical parameters of heat exchanger take either lower or upper limit values from their ranges of validity mentioned in Table 1. This restricts the process of selection of unique set of heat exchanger parameters. The observed behavior is due to the absence of peak in the j/f or Nu/f profile. Consequently, results from an alternative method are described in the following sub-sections.

Table 2 j_{max} and f_{min} from various correlations

	Webb and Gray [11]	Kim et al. [10]	Wang et al. [12]	Xie et al. [13]	Wang et al. [14]
j_{max}	0.0278	0.0304	0.1541	0.4316 (Nu_{max})	0.052
f_{min}	0.3388	0.1786	0.0617	2.001	0.0593

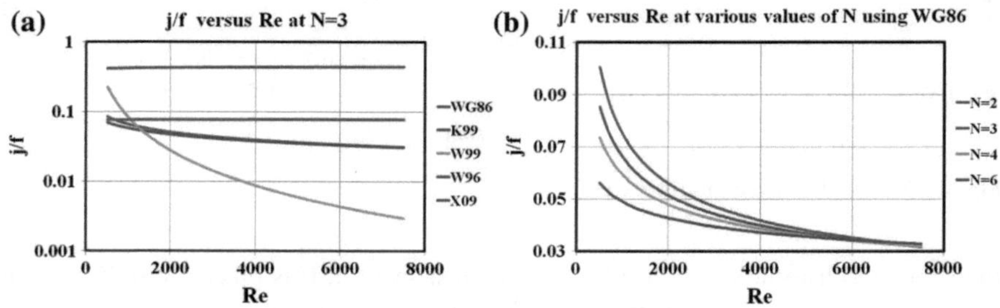

Fig. 5 **a** Comparison of various correlations for j/f. **b** Comparison of j/f for various number of tube rows

3.1 *Comparison of Correlations and Number of Tube Rows*

A comparison of various correlations (see Table 1) is shown in Fig. 5a. It can be seen that the correlation with very large and very low j/f shows extreme behavior in the group. It is natural, therefore, to go for the correlation which gives higher j/f ratio. However, on a conservative side, a correlation of Webb and Gray [11] and Xie et al. [13] seems to show acceptable behavior of lesser variance in the group. Moreover, among these two correlations, Webb and Gray [11] is found to be the best in regards to low RMS error, better range of validity, consistency and considerable conservation in design. Hence, correlation of Webb and Gray [11] is selected for further study. A comparison of Reynolds number (*Re*), with the figure of merit (j/f) for various values of number of tube rows (*N*), is shown in Fig. 5b. It can be seen that for $Re > 6000$, number of tube rows has negligible effect on j/f. On the other hand, for low *Re* (<3000) the effect is pronounced. Moreover, it can be seen that the difference of j/f between two-tube-rows and three-tube-rows is high compared to that among three- and four-tube-rows. Similarly, difference of j/f between four-tube and six-tube-rows is high. Considering these aspects, three-tube rows are recommended as it gives a conservative variation of the figure of merit at all applied conditions.

3.2 Selection of Fin Type and Qualitative Comparison of Wet and Dry Air

Figure 6a shows comparison of experimental data of Kang et al. [17] for Nu/f for different types of fins, i.e., plain fin (PF), wavy fin (WF), and slit fin (SF) with different fin spacing for $N = 3$. For the comparison of Nu/f, fitted correlations of $Nu = F(Re)$ and $f = F(Re)$ to the experimental data are used. From Fig. 6a, it is observed that the slit fins with $P_f/D = 0.197$ give the best result, followed by plain fins with $P_f/D = 0.256$ and 0.315. The wavy fins (triangular and sinusoidal), for all P_f/D values, and slit fins with $P_f/D > 0.256$ show comparable *Nu/f* with plain fins. However, considering the ease of manufacturing and cost saving as the additional benefits, the plain fins are selected for the future study. In Fig. 6a, furthermore, numerical data of Yin et al. [18] for plain fins with $P_f/D = 0.256$ is compared. It shows that the deviation in experimental and numerical data at $Re > 10{,}200$ is in between 11 and 25%. The deviation in the data at higher *Re* may be related to the complexity of the flow, which could not be properly captured by the steady and laminar numerical model. However, the deviation is less than 1% for $Re < 6000$ which is well within the range of present work. Thus, with the additional consideration of ease in manufacturing, plain fins are suggested.

Till now the correlations available for dry air are compared to decide on design parameters of AWHX as a initial tool for its design. It would be of interest to qualitatively compare the wet-air correlations with the dry air. Therefore, Fig. 6b shows comparison of *j* at various *Re* for wet-air (with 80% *RH*) and dry air. It is interesting to see that the wet air shows improvement in the heat transfer factors by 10%.

From the above results and discussions, therefore, the heat exchanger (AWHX) consists of plain fin and three-tube-rows configuration with various parameters, as mentioned in Table 3, are suggested for future wet-air side experiments. For these experiments, the wind-tunnel dimensions, which are shown in Table 4, are obtained.

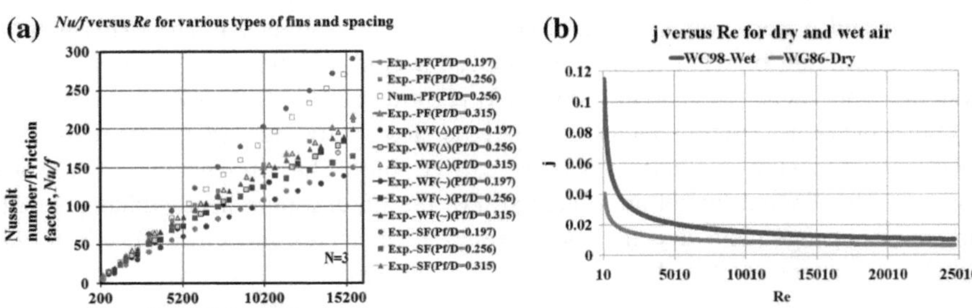

Fig. 6 **a** Comparison of Nu/f for various fin types [*PF* plain fin, *WF* wavy fin (Δ: triangular and ~: sinusoidal), *SF* split fin, Exp: Kang et al. [17], Num: Yin et al. [18]]. **b** Comparison of *j* for wet (80% RH) and dry air

Table 3 Dimensions of heat exchanger

Parameters	Values
Fin type and tube arrangement	Plain and staggered
RH	45–85%
RE	500–6000
N	3
D	5, 8, 12.5 mm
P_f/D	0.65, 2.92, 4.7
P_t/D	2.0, 3.5, 5.0
P_t/P_l	1
P_l/D	2.0, 3.5, 5.0

Table 4 Dimensions of wind tunnel for maximum 10 m/s velocity inside test section

	L (m)	W (m)	H (m)	α or β (°)	Diameter (m)	Pressure drop with wet-air (Pa)	Power (HP) and cfm
Settling chamber	0.54	0.9	0.8	–	–	27	–
Contraction cone	1.4	0.9, inlet 0.3, outlet	0.8, inlet 0.2, outlet	12°		2	–
Test section	3.00	0.3	0.2	–	–	458	–
Diffuser	1.5	0.3, inlet 0.48, outlet	0.2, inlet 0.38, outlet	3.5°		3	–
Fan/blower	0.20	–	–	–	0.4	Total, $\geq$ 490 Pa ($\geq$ 2 in Wg)	$\geq$ 1 HP and 1500 cfm

4 Conclusions

In this paper an experiment design for hybrid condenser is presented. Considering scarcity of water and lower ITD in Rajasthan, India, use of earth/ground cooled water/air and water spray cooling of condenser air is suggested. The experimental design consisting of wind tunnel, test-section accommodating water spray cooling system, and heat exchanger, considering the ratio of Nusselt number or Colburn-factor to friction factor for various flow velocities, and state-of-the art wind-tunnel design methodologies, is carried out. The dimensions of various components of wind tunnel are obtained. For the fin-and-tube heat exchanger with staggered arrangement, dry-air side correlation of Webb and Gray [11] is found suitable in selecting AWHX parameters for the experiment design. Hence, heat exchanger consists of plain fin and three-tube-rows configuration with $P_f/D = 0.256$ or 0.315, $P_t/D = P_l/D = 2.0$, and $\delta_f/D = 0.02$, along with the values higher and lower than these, are suggested for future wet-air side experiments in the range $500 < Re < 6000$ (based on the heat exchanger tube outer diameter).

References

1. J.G. Bustamante, A.S. Rattner, S. Garimella, Achieving near-water-cooled power plant performance with air-cooled condensers. Appl. Therm. Eng. xxx:1–10, (2015) http://dx.doi.org/10.1016/j.applthermaleng.2015.05.065
2. A.M. Blanco-Marigorta, M.V. Sanchez-Henríquez, J.A. Peña-Quintana, Exergetic comparison of two different cooling technologies for the power cycle of a thermal power plant. Energy **36**, 1966–1972 (2011)
3. H.L. Zhang, J. Baeyens, J. Degŕeve, G. Cacères, Concentrated solar power plants: review and design methodology. Renew. Sustain. Energy Rev. **22**, 466–481 (2013)
4. C. Kutscher, D. Costenaro, Assessment of evaporative cooling enhancement methods for air-cooled geothermal power plants. Geothermal Resources Council (GRC) Annual Meeting Reno, Nevada September 22–25 (2002)
5. A. Alkhedhair, Z. Guan, I. Jahn, H. Gurgenci, S. He, Water spray for pre-cooling of inlet air for natural draft dry cooling towers—experimental study. Int. J. Therm. Sci. **90**, 70–78 (2015)
6. J. Tissot, P. Boulet, A. Labergue, G. Castanet, F. Trinquet, L. Fournaison, Experimental study on air cooling by spray in the upstream flow of a heat exchanger. Int. J. Therm. Sci. **60**, 23–31 (2012)
7. S.A. El-Agouz, A.E. Kabeel, Performance of desiccant air conditioning system with geothermal energy under different climatic conditions. Energy Convers. Manage. **88**, 464–475 (2014)
8. L. Chandra, A. Agarwal, R.V. Maitri, P. Garg, Design and analyses of earth-air heat exchange systems for space cooling, *IEEE–Third International Conference on Sustainable Energy Technologies (ICSET)*, Kathmandu, Nepal, pp 385–390 (2012) https://dx.doi.org/10.1109/ICSET.2012.6357430
9. C.C. Wang, C.T. Chang, Heat and mass transfer for plate fin-and-tube heat exchangers, with and without hydrophilic coating. Int. J Heat Mass Transf. **41**, 3109–3120 (1998)
10. N.H. Kim, B. Youn, R.L. Webb, Air-side heat transfer and friction correlations for plain fin-and-tube heat exchangers with staggered tube arrangements. Trans. ASME **221**, 662–667 (1999)
11. R.L. Webb, D.L. Gray, Heat transfer and friction correlations for plate finned tube heat exchangers having plain fins, in *Proceedings of the 8th Heat Transfer Conference*, pp 2745–2750 (1986)
12. C.C. Wang, K.Y. Chi, C.J. Chang, Heat transfer and friction characteristics of plain fin-and-tube heat exchangers, part II: correlation. Int. J. Heat Mass Transf. **43**, 2693–2700 (1999)
13. G. Xie, Q. Wang, B. Sunden, Parametric study and multiple correlations on air-side heat transfer and friction characteristics of fin-and-tube heat exchangers with large number of large-diameter tube rows. App. Therm. Engg. **29**, 1–16 (2009)
14. C.C. Wang, Y.J. Chang, Y.C. Hseih, Y.T. Lin, Sensible heat and friction characteristics of plate fin-and-tube heat exchangers having plane fins. Int. J. Refrig. **19**(4), 223–230 (1996)
15. X. Ma, G. Ding, Y. Zhang, K. Wang, Airside heat transfer and friction characteristics for enhanced fin-and-tube heat exchanger with hydrophilic coating under wet conditions. Int. J. Refrig. **30**, 1153–1167 (2007)

16. R. Yun, Y. Kim, Y. Kim, Air side heat transfer characteristics of plate finned tube heat exchangers with slit fin configuration under wet conditions. Appl. Therm. Eng. **29**, 3014–3020 (2009)
17. H.J. Kang, W. Li, H.Z. Li, R.C. Xim, W.Q. Tao, Experimental study on heat transfer and pressure drop characteristics of four types of plate fin-and-tube heat exchanger surfaces. J. Therm. Sci. **3**(1), 34–42 (1994)
18. J. Yin, Z. He, F. Chen, J. Ma, Effect of tube location change on heat transfer characteristics of plain plate fin-and-tube heat exchangers. J. Therm. Sci. Eng. Appl. **6**(021005), 1–9 (2014)
19. M.A. González Hernández, A.I. López, A.A. Jarzabek, J.M. Perales Perales, Y. Wu, S. Xiaoxiao, Chapter 1: design methodology for a quick and low-cost wind tunnel, in *Wind tunnel designs and their diverse engineering applications*, ed by N.A. Ahmed (InTech Publisher, 2013)
20. S. Brusca, R. Lanzafame, M. Messina, Chapter 7: low-speed wind tunnel: design and build, in wind tunnel aerodynamics, models and experiments, in *Wind tunnels: aerodynamics, models and experiments*, ed. by J. Pereira (Nova Science Publications Inc., New York, 2011)
21. J.B. Barlow, W.H. Rae, A. Pope, *Low-Speed Wind Tunnel Testing*, 3rd edn. (Wiley, London, 1999)
22. G. Biswas, K. Torii, D. Fuji, K. Nishino, Numerical and experimental determination of flow structure and heat transfer effects of longitudinal vortices in a channel flow. Int. J. Heat Mass Transf. **39**(16), 3441–3451 (1996)
23. H.T. Chen, J.R. Lai, Study of heat-transfer characteristics on the fin of two-row plate finned-tube heat exchangers. Int. J. Heat Mass Transf. **55**, 4088–4095 (2012)

Experimental and Computational Investigation of Heat Transfer in an Open Volumetric Air Receiver for Process Heat Application

P. Sharma, Laltu Chandra, Rajiv Shekhar and P.S. Ghoshdastidar

Abstract India receives abundant solar irradiance with an annual average of ~ 19.97 MJ/m^2 per day in Jodhpur only. This solar energy can be harnessed for electricity generation, melting or heat treatment of metals. Use of air as heat transfer fluid offers significant advantages of being nontoxic, freely available and operating temperature beyond 800 °C. Considering these aspects, as a research initiative, open volumetric air receiver (OVAR) is being developed with a peak-power capacity of 4 kW$_{th}$. The installed testing facility at IIT Jodhpur includes sub systems, which are thermal energy storage (TES), air–water heat exchanger. In the absence of solar simulator electrical heating is being employed for circumferential (external) heating of the absorbers. In particular, the presented paper presents:

(a) Effect of pore diameters (2 and 3 mm) on the average outlet temperature of absorber with porosity (ε) $\sim 52\%$ at POA/MFR = 100 kJ/kgK, where POA is the equivalent Power-On-Aperture and MFR is mass-flow rate of air;
(b) Efficiency performance curve for absorbers with $\varepsilon \sim 52\%$;
(c) Modeling of heat transfer in absorber with adopted commercial CFD tool FLUENT including returned air circulation;
(d) Comparison between CFD analyzed and experimentally obtained temperature for absorbers with $\varepsilon \sim 42$, 52, and 62%;
(e) Predictions with incident radiation onto the front surface of porous absorber.

P.S. Ghoshdastidar
Department of Mechanical Engineering, Indian Institute of Technology Kanpur, Kanpur 208016, India

L. Chandra (✉)
Department of Mechanical Engineering and Center for Solar Energy, Indian Institute of Technology Jodhpur, Jodhpur 342011, Rajasthan, India
e-mail: chandra@iitj.ac.in

P. Sharma · R. Shekhar
Department of Materials Science and Engineering, Indian Institute of Technology Kanpur, Kanpur 208016, India

L. Chandra and A. Dixit (eds.), *Concentrated Solar Thermal Energy Technologies*, Springer Proceedings in Energy, https://doi.org/10.1007/978-981-10-4576-9_10

Keywords Solar energy · Open volumetric air receiver (OVAR) · Absorber · CFD · POA/MFR · Local thermal nonequilibrium model (LTNE)

Nomenclature

ρ_f	Density of fluid (kg/m^3)
$v(v_1, v_2, v_3)$	Velocity vector
E_f	Total fluid energy (J)
η_{th}	Thermal efficiency
S_s^h	Solid enthalpy source term
T_f	Temperature of fluid (K)
t	Time (s)
μ	Viscosity of air (kg/m s)
k_f	Thermal conductivity of fluid (W/mK)
RANS	Reynolds Average Navier–Stokes
h_{fs}	Heat transfer coefficient for the fluid/solid interface
T_s	Temperature of solid (K)
ε	Porosity (%)
p	Static pressure (Pa)
y^+	Nondimensional wall distance
S_f^h	Fluid enthalpy source term
A_{fs}	Interfacial area density, i.e., the ratio of the area of the fluid/solid interface and the volume of the porous zone

1 Introduction

Growing energy demands and environmental impact of fossil fuels needs to look for the alternative fuel sources mainly from the renewable sources [1]. Solar energy is acknowledged as a prominent alternative. To harness solar energy, in an efficient manner, optical concentration is required to achieve high temperatures. There are various methods available for concentrating solar irradiance, like line focusing techniques, which include parabolic trough and linear Fresnel technology and point focusing techniques, which include dish Stirling and heliostats-based solar tower [2]. Solar tower uses heliostats that focus solar radiation onto the front surface of the absorbers. This results in volumetric heating of porous absorbers. Ambient air from the atmosphere is sucked through the porous absorbers, which in turn removes the heat from absorbers and, thus, relatively hot air exits the receiver assembly [3]. This technology can achieve temperature in the range of 900–1100 °C [4]. Since this technology is under research from 1980's and several kind of absorbers are tested and demonstrated successfully, which makes it key contender to use at higher temperature levels [5]. This study focuses on experimental and computation work carried out using a receiver assembly. Absorbers with pore diameters of 2 and 3 mm for a fixed porosity of $\sim$52% were evaluated experimentally. The efficiency or performance curves are presented for the selected absorbers at various $\frac{\text{POA}}{\text{MFR}}$ keeping the POA as constant. Finally, the adopted Local Thermal Nonequilibrium model (LTNE) as in ANSYS Fluent for modeling the heat transfer in porous absorber is validated using the measured data.

2 Experiment: Setup and Measurement

The schematic of the experimental setup is shown in Fig. 1. There are seven absorbers marked A are placed inside steel casing with an inner diameter of ~100 mm. The brass absorbers are cylindrical in shape having pore diameter of 2 and 3 mm with porosity ~52%. The smallest gap between absorbers is 2.54 mm with diameter of the absorber as ~25.4 mm. Cold air is sucked with the help of a blower through the absorbers, which is heated inside the absorbers with electrical heating. Subsequently, the obtained hot air from the receiver enters the installed TES in which heat is transferred to the rock/pebbles. Finally, the relatively cold air from TES enters the rotameter, which is also calibrated up to 60 °C. As per ongoing measurements an uncertainty of ±5% is expected. The cold air is returned through an insulated pipe to receiver wherein it impinges on the footpieces. Other details of the experimental setup are presented with Sharma et al. [6].

2.1 *Effect of Pore Diameter*

As explained, absorbers with pore diameter of 2 and 3 mm were tested keeping the porosity ~52%. For this purpose, central pore temperatures (r = 0) are measured using thermocouples at the outlet of the absorbers, which are installed at 30 mm from the inlet. In this paper central pore temperature in the absorbers 1 and 3 are compared. Further, averaged values of measured pore temperature at r = 0 over the seven absorbers are compared at various POA with POA/MFR ~ 100.

Figure 2b shows that the average pore temperature at r = 0 in case of absorber with pore diameter of 3 mm is higher than that of 2 mm. This is attributed to decrease in velocity of air inside the absorber pores for the same mass-flow rate and

(a) (b)

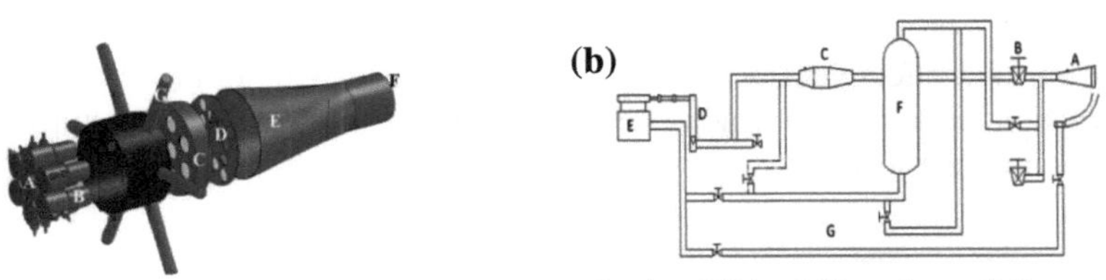

A: Absorbers; **B**: Footpieces; **C**: Anchor plate; **D**: Perforated plate; **E**: Mixer; **F**: Receiver outlet; **G**: Injection port

A: Receiver, **B**:Valve, **C**: Heatexchanger, **D**: Rotameter, **E**: Blower, **F**:Thermal Energy Storage, **G**:Return air line

Fig. 1 **a** Exploded view of the receiver assembly, **b** schematic of the experimental setup

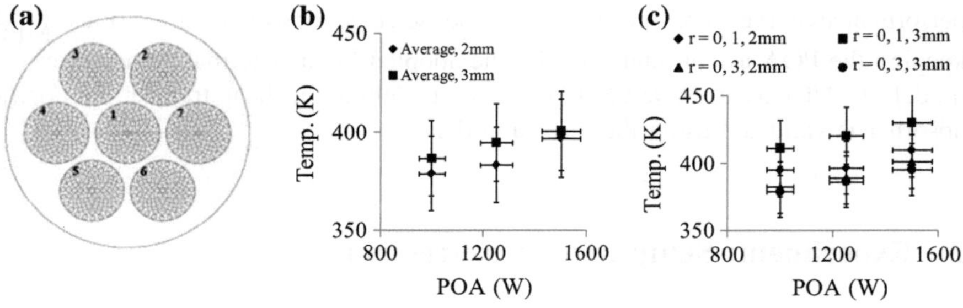

Fig. 2 **a** Nomenclature of the absorbers; **b** comparison of average temperature, **c** pore temperature at $r = 0$ of absorber with absorber porosity of $\sim 52\%$

a comparable porosity. As expected, the average and local measurement shows the same behavior for various POA values in Fig. 2c. Thus, larger pore diameter may be preferred for durability. However, for achieving higher temperature, a smaller pore diameter may be recommended.

2.2 Absorber Performance

This section deals with the efficiency or performance of the selected absorbers having porosity $\sim 52\%$ for different POA/MFR values and POA of 750 W. Joule or resistance heating is used for volumetric heating of absorbers. This is demonstrated in Sharma et al. [6]. For this purpose, Nichrome wire is circumferentially wrapped around the absorbers. The input voltage is controlled using four variacs. The resulting current is measured with the help of ammeter. Thus, the power applied is calculated as:

$$P = I^2 R \tag{1}$$

P Power to absorber (W)
I Current to absorber (A)
R Resistance of Nichrome wire (Ω)

The performance of the experimental absorbers is depicted in Fig. 3. This clearly shows that with increase of POA/MFR the outlet air temperature increases and the heat transfer effectiveness decreases. This is attributed to increase in—(a) the convection and radiation heat losses and (b) the rate of heat gain by return air a part of which is lost to ambient.

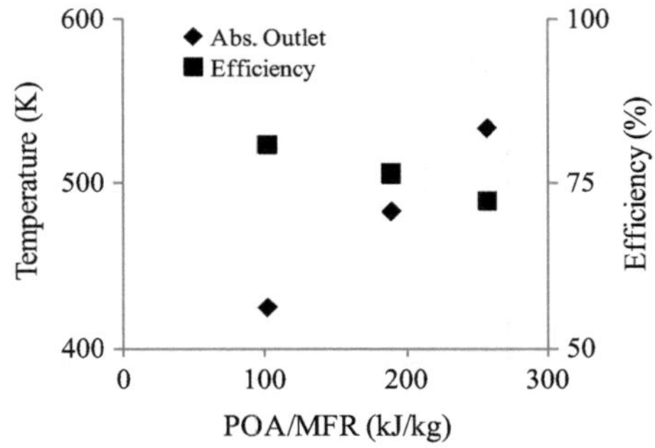

Fig. 3 Effect of POA/MFR on outlet temperature and efficiency of the absorber

3 Simulation: RANS-Based CFD Simulations

3.1 Governing Equations

The employed RANS-based CFD approach solves the following equations in their time averaged form [ANSYS Fluent 12.0]:

Continuity:

$$\frac{\partial \rho_{\mathrm{f}}}{\partial t} + \nabla \cdot (\varepsilon \rho_{\mathrm{f}} v) = 0$$

Momentum:

$$\frac{\partial}{\partial t}(\rho v) + \mathrm{div}(\rho v v) = \mathrm{div}(\mu \mathrm{grad} v) - \frac{\partial p}{\partial j} + B_j + v_j$$

The LTNE model solves energy equations for fluid and solid separately. The energy equations for fluid and solid are as follows:

$$\text{Energy: fluid} \quad \frac{\partial}{\partial t}(\varepsilon \rho_{\mathrm{f}} E_{\mathrm{f}}) + \nabla \cdot (\vec{v}(\rho_{\mathrm{f}} E_{\mathrm{f}} + p)) = \nabla \cdot \left(\varepsilon k_{\mathrm{f}} \nabla T_{\mathrm{f}} - \left(\sum i h_i J_i\right) + (\bar{\bar{\tau}}.\vec{v})\right) + S_{\mathrm{f}}^{\mathrm{h}} + h_{\mathrm{fs}} A_{\mathrm{fs}} (T_{\mathrm{f}} - T_{\mathrm{s}})$$

$$\text{Energy: solid} \quad \frac{\partial}{\partial t}((1-\varepsilon)\rho_{\mathrm{s}} E_{\mathrm{s}}) = \nabla \cdot ((1-\varepsilon) k_{\mathrm{s}} \nabla T_{\mathrm{s}}) + S_{\mathrm{s}}^{\mathrm{h}} + h_{\mathrm{fs}} A_{\mathrm{fs}} (T_{\mathrm{s}} - T_{\mathrm{f}})$$

3.2 Solution Methodology

In lieu of solar simulator, which focuses the solar radiation on the front surface of the absorbers, electrical heating is employed in the presented experiment.

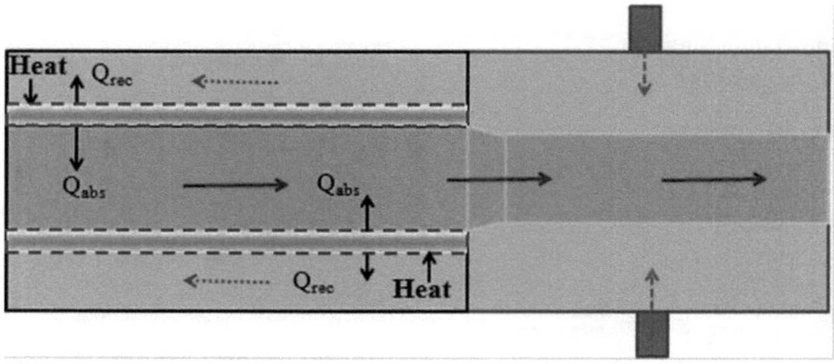

Fig. 4 Schematic of the absorber domain

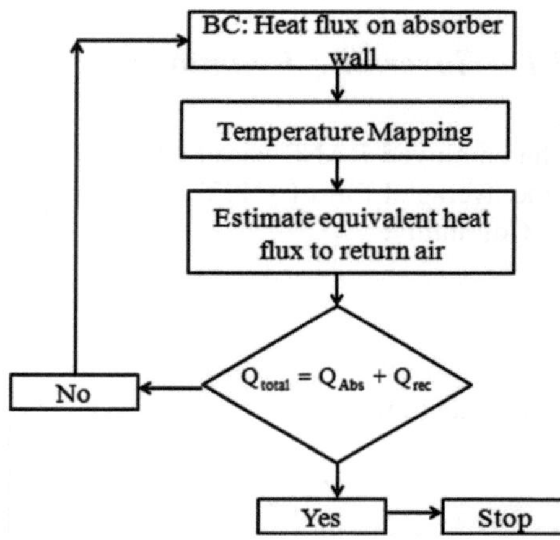

Fig. 5 Flow chart of the adopted solution methodology

As described, the Local Thermal Nonequilibrium (LTNE) model is used in which temperature of solid is not considered as equal to the fluid temperature. Instead of modeling the absorber as manufactured, the pores of the absorbers were replaced with the equivalent porosity. Due to the use of LTNE model the circumferential wall of the absorber breaks into two distinct walls. This allows fractional transfer of the applied heat to air flowing in the pores and the rest to return air flowing around the absorbers. To ensure equal temperature on the shared-wall an UDF is implemented in the FLUENT. The wall-sharing methodology is depicted in Fig. 4.

The flow chart of implemented solution methodology is shown in Fig. 5. Here, heat flux is applied on the circumferential walls of the absorbers to simulate the performed experiments.

The schematic diagram of the computational domain with boundary conditions is shown in Fig. 6 at the inlet of the absorbers. Due to geometrical symmetry one-quarter of the experimental absorber assembly with foot-piece is considered for

Fig. 6 The schematic diagram and applied boundary conditions of the simulated domain at the inlet

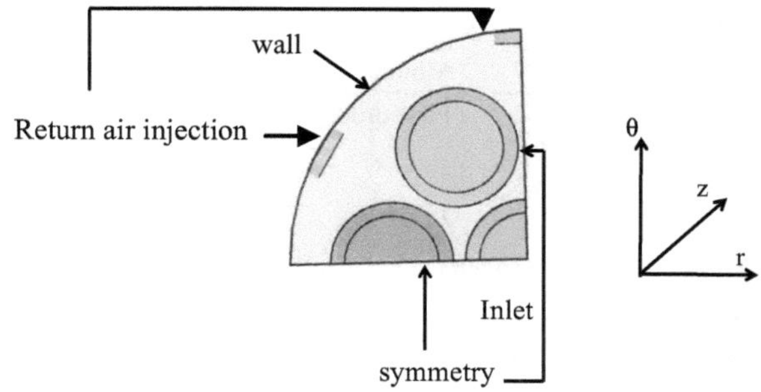

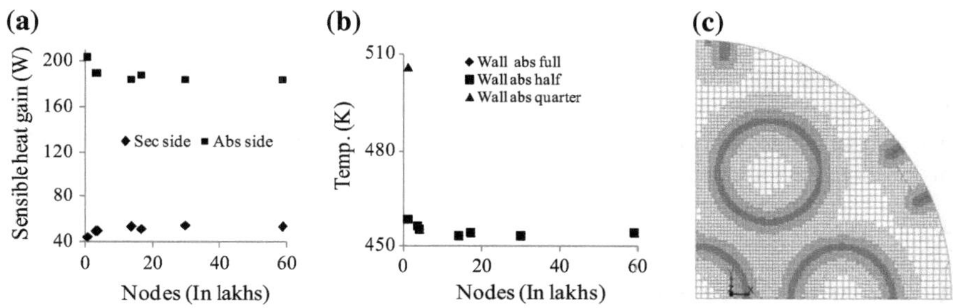

Fig. 7 **a** Grid independence test for heat transfer rate by primary and secondary air; **b** circumferential wall of the absorber; **c** mesh adopted for the validation purpose

modeling. This also includes one and half of the return-air injection ports. The following boundary conditions were used for the numerical analysis:

(a) $z = 0$ (inlet to absorber and outlet of return air): $\dot{m}_{\rm f} = C1$ and $T_{\rm f} = C2$, $P = P_{\rm atm}$
(b) $z = L$ (outlet): $\frac{\partial T}{\partial Z} = 0, P = P_{\rm atm}$
(c) $r = R(z)$: $\left.\frac{\partial T}{\partial r}\right|_{r=R(z)} = 0$ (adiabatic wall) and $u = v = w = 0$.
(d) $z = 42.42$ mm and $\theta = 0°$ and $45°$: $\dot{m}_{\rm f} = C1$ and $T_{\rm f} = C3$
(e) $r = R$(Absorber wall): $q'' = \text{constant}$

3.3 Grid Independence Test

For selecting the final mesh with simulation domain various resolutions are employed. It is observed that with the number of computational nodes beyond 1.4 Million there is no practical change in the heat transfer to primary and return air from absorber surface. Similarly behavior is obtained for circumferential wall temperature of the selected absorber as in Fig. 7. Therefore, the final mesh resolution with 1.4 Million of computational nodes is selected.

Table 1 The adopted numerical setup for RANS-based CFD analysis

Mesh	Momentum and energy	Turbulence model	Convergence
$Y^{+} \sim 7$ Aspect ratio = 26	First order upwind	*k*-epsilon	10^{-4} to 10^{-7}

Table 2 Cases considered for the validation

Porosity (%)	Power-on-aperture (W)	Mass-flow rate (kg/s)	POA/MFR (kJ/kg)
42	1000	6.9×10^{-3}	145
52	1000	10.0×10^{-3}	100
62	1000	6.9×10^{-3}	145

The employed mesh and numerical scheme for Reynolds Averaged Navier Stokes (RANS)-based CFD model and the adopted convergence criteria are summarized in Table 1. The level of convergence provides confidence on the stability of the obtained numerical data.

4 Results: Validation and Application

This section deals with the validation of the adopted numerical setup and its application for further analysis. The experimental conditions are summarized in Table 2. This shows the selected absorbers with different porosities, namely, 42, 52, and 62% are subject to various POA and POA/MFR conditions. The same POA/MFR for absorbers with 42 and 62% porosities is considered for completeness. This will allow comparing the separate effect of porosity in addition to integrated effect.

The estimation of heat transfer coefficient to return air (h_{sf}) is required for using LTNE model in FLUENT as input. This is in addition to the described boundary conditions. For this purpose a correlation which is applicable to the considered problem is selected from Xu et al. [7] and is as follows:

$$h_{\mathrm{sf}} = 1.064\left(\frac{k_{\mathrm{f}}}{d_{\mathrm{p}}}\right)\mathrm{Pr}^{0.33}\mathrm{Re}_{d}^{0.59} \quad \text{for } \mathrm{Re}_{d} \geq 350 \tag{1}$$

Figure 8a shows the comparison of experimentally and numerically analyzed heat transfer rate (power) to primary air in the pores and return air around the absorbers. A deviation of less than 5% is observed between experiments and simulated values. Moreover, the measured and calculated temperatures at the absorber and return-air outlet are compared in Fig. 8b. This also depicts that the adopted CFD method is capable of predicting within an uncertainty of about 5%.

Table 3 compares experimentally and numerically obtained values of (a) outlet air temperature of absorber and return air and (b) heat transfer rate to primary and

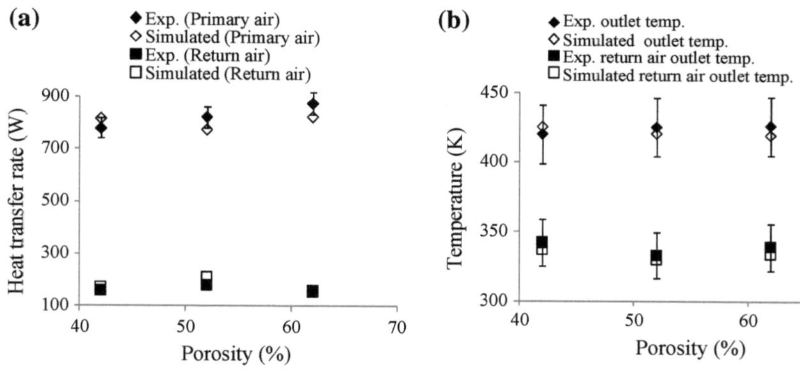

Fig. 8 Comparison of experimental and simulated: **a** heat transfer rate to primary and return air, **b** absorber and return-air outlet temperature

return air. This clearly demonstrates that the adopted modeling strategy is acceptable and captures the involved heat transfer process in porous absorber. The total input power is estimated as:

$$\text{Total input power} = \text{Power to (primary} + \text{return} + \text{footpiece)}$$

After validation was performed and wall resolved mesh is obtained, predictive analyses are performed by applying the heat flux from the front surface of the absorbers. In these simulations, uniform heat flux is imposed on the front surface of absorbers. The penetration profile of solar irradiance along the absorber depth was ignored for this prediction. Both convective and radiative losses are considered from the front surface of absorber to ambient for calculating the outlet temperature and the respective efficiency of the porous absorber. The absorbers are considered as gray body having emissivity of 0.8 and heat transfer coefficient is obtained from Zheng et al. [8]. The conditions are summarized in Table 4 to analyzed the effect of POA and POA/MFR, separately.

The numerically obtained values are represented in Fig. 9. This shows the following:

(a) the outlet temperature of absorber increases with POA/MFR for a fixed POA value;
(b) with increasing air temperature absorber efficiency decreases as a result of losses to ambient and heat transfer to return air from absorber;
(c) finally, Fig. 9b shows that outlet temperature increases with POA for a fixed value of POA/MFR. It may be concluded safely, the absorber efficiency/performance will deteriorate with increasing POA in such a condition.

Table 3 Comparison between experiment and numerical values of various parameters

Porosity	h_{sf} (W/m^2 K)	Exp. heat transfer rate by primary air (W)	Simulated heat transfer rate by primary air (W)	Exp. heat transfer rate by return air (W)	Simulated heat transfer rate by return air (W)	Exp. absorber outlet temperature (K)	Simulated absorber outlet temperature (K)	Exp. return-air outlet temperature (K)	Simulated return-air outlet temperature (K)
42	485	780	816	160	172	420	425	342	337
52	540	824	775	183	212	425	420	333	330
62	384	877	823	158	160	426	419	339	334

Table 4 Predicted effect of POA/MFR and POA on the performance of absorber

POA/MFR (kJ/kg), [POA = 1500 W]	MFR × 10^{-3} (kg/s)	POA (W) [POA/MFR = 300 kJ/kg]	MFR × 10^{-3} (kg/s)
100	15	1000	3.33
200	7.5	1250	4.16
300	5	1500	5

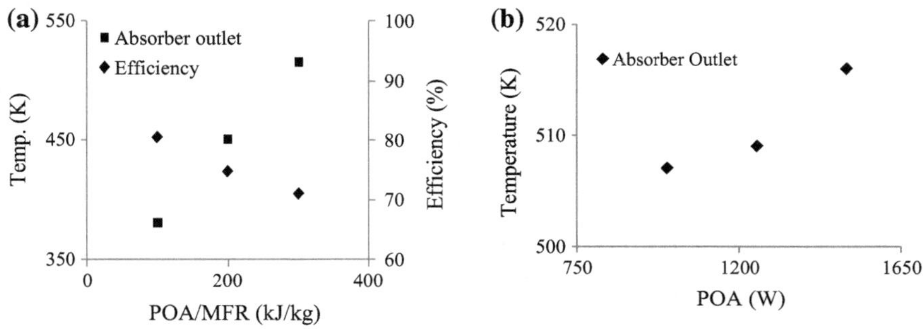

Fig. 9 **a** Effect of $\frac{\text{POA}}{\text{MFR}}$ on the outlet absorber temperature and efficiency, **b** Effect of POA on the absorber outlet temperature

5 Conclusion

This paper presented an experimental analysis of absorber keeping the porosity of 52% and varying the pore diameter from 2 to 3 mm. There is not much increase in the temperature of absorber with pore diameter 3 mm. Effect of POA/MFR is shown for a constant power of 1250 W by varying the ratio of POA/MFR from 100 to 400. There is a drop in the efficiency with an increase in the temperature of the air of absorber outlet. Several experiments were performed with varying the porosity of the absorbers and at various POA and POA/MFR. Three experiments were chosen with varying porosity, POA and POA/MFR for the validation purpose of the CFD tool FLUENT. It can be seen that in the sensible air heat gain by the return air and primary air is within 5% accuracy. In case of outlet temperature from absorber and return air the deviation is less than 5%. For predicting the effect of heat flux from the front surface of the absorber, heat flux is impinging from the front surface of the absorbers as in the real situation where heliostats focuses the solar radiation to the absorber placed at the top of a tower. This also shows that there is a rise in temperature and drops in efficiency as the ratio of POA/MFR is increasing keeping the constant POA of 1500 W, further there is a rise in temperature as POA on the front surface is increasing keeping the ratio of POA/MFR fixed as 300.

References

1. R. Capuano, T. Fend, P. Schwarzbözl, O. Smirnova, H. Stadler, B. Hoffschmidt, R. Pitz-Paal, Numerical models of advanced ceramic absorbers for volumetric solar receivers. Renew. Sustain. Energy Rev. **58**, 656–665 (2016)
2. R. Sarma, P.K. Sharma, D. Kumar, L. Chandra, P.P. Kumar, R. Shekhar, On the applications of heat and fluid flow for designing solar thermal sub-systems. In: Proceedings of the ICORE 2012: International Congress on Renewable Energy, pp. 219–229 (2012)
3. B. Hoffschmidt, F.M. Téllez, A. Valverde, J. Fernández, V. Fernández, Performance evaluation of the 200-kW_{th} HiTRec-II open volumetric air receiver. J. Sol. Energy. Eng. **125**, 87–94 (2003)
4. A.L. Avila-Marín, Volumetric receivers in solar thermal power plants with central receiver system technology: a review. Sol. Energy **85**, 891–910 (2011)
5. C.K. Ho, B.D. Iverson, Review of high-temperature central receiver designs for concentrating solar power. Renew. Sustain. Energy Rev. **29**, 835–846 (2014)
6. P. Sharma, R. Sarma, L. Chandra, R. Shekhar, P.S. Ghoshdastidar, Solar tower based aluminum heat treatment system: Part I. Design and evaluation of an open volumetric air receiver. Sol. Energy **111**, 135–150 (2015)
7. C. Xu, Z. Song, L. Chen, Y. Zhen, Numerical investigation on porous media heat transfer in a solar tower receiver. Renew. Energy **36**, 1138–1144 (2011)
8. H. Zheng, X. Yu, Y. Su, S. Riffat, J. Xiong, Thermodynamic analysis of an idealised solar tower thermal power plant. Appl. Therm. Eng. **81**, 271–278 (2015)

Experimental Performance Evaluation of a Parabolic Solar Dish Collector with Nanofluid

Jayesh Novel Pakhare, Harikesh Pandey, Mari Selvam and C.P. Jawahar

Abstract The experimental performance of a parabolic solar dish collector is investigated in this paper. The collector is located at Karunya University, Coimbatore (Longitude 76° 59′ and Latitude 11° 00′) Tamil Nadu. The diameter of the collector is 1180 mm with the depth of 310 mm. The diameter of the absorber which is made up of copper is about 75 mm and height is 200 mm for a concentration ratio of 10. The performance of the collector is evaluated without and with nanoparticles, under various operating conditions. It is observed that, without nanoparticles, the temperature of the water in the storage tank increased from 35 to 85 °C. When aluminium oxide nanoparticles (0.01% volume concentration) are used, the temperature increased up to 90 °C. The efficiency of the collector is found to be higher by about 10% when nanoparticles are used.

Keywords Aluminium oxide · Nanofluid · Parabolic collector · Volume concentration

1 Introduction

As natural resources are decreasing day by day we need some sources of energy as substitute which are energy efficient, pollution free and economical. Solar energy is one of these. Solar energy can be used for heating purpose at micro and macro level. When solar energy is used in heat generation systems like solar dish type

J.N. Pakhare (✉) · H. Pandey · M. Selvam · C.P. Jawahar
Karunya University, Coimbatore, Tamil Nadu, India
e-mail: jayesh.pakhare@gmail.com

H. Pandey
e-mail: akashpandey790@gmail.com

M. Selvam
e-mail: mariselvama11feb@gmail.com

C.P. Jawahar
e-mail: jawahar@karunya.edu

L. Chandra and A. Dixit (eds.), *Concentrated Solar Thermal Energy Technologies*, Springer Proceedings in Energy, https://doi.org/10.1007/978-981-10-4576-9_11

collectors, the efficiency of the systems can further be increased by using nanofluids for heat transfer and optical absorption. Higher optical efficiency and thermal efficiency can be obtained by the parabolic solar dish collectors. A parabolic-shaped dish concentrator that reflects solar radiation onto a receiver is mounted at the focal point of the parabolic dish system. The heat collected by the receiver can be used by a heat engine mounted on the receiver that moves with the dish structure, such as Stirling and Brayton cycle engines. The shape of the parabola needs to be relatively precise and concentrator needs to focus the sunlight on the receiver in order for the parabolic concentrator to be most effective. The utilization of solar energy by converting it to heat is the principal concern of the solar thermal technology. Solar energy is collected and concentrated so that higher temperatures can be obtained in the concentrating type of solar collector. The surface temperature of the sun is the limit. The solar radiant energy can be more efficiently harnessed by the Nanofluid-based solar collector compared to conventional solar collectors. Relevant application of Nanofluids have recently found in industrial applications, applications requiring quick and effective heat transfer such as, cooling of microchips, microscopic fluidic applications, etc.

2 Literature Survey

The use of reflectors [1] to concentrate sunlight on the absorber of a solar collector, reduces the absorber size, which in turn reduces heat losses and dramatically increases efficiency at high temperatures. The cost of reflectors is significantly less per unit area than collectors, which is yet an added advantage.

Basic types of concentrating collectors are as follows:

1. Parabolic trough
2. Parabolic dish
3. Stationary concentrating collectors

Li and Dubinsky [2] have worked on a new approach to design a concentrating dish using flexible petals. In solar collectors, parabolic dish concentrator mirrors are very important parts as these are used to concentrate the sun rays to the focal point of the parabola and they should be relatively precise and are expensive to fabricate and also to transport. Wua et al. [3] proposed a solar thermal power system of parabolic dish/AMTEC. They worked on a combined system in which a parabolic dish solar collector was combined through a coupling heat exchanger with an alkali metal thermal to electric converter (AMTEC). Its overall performance of thermo-electric conversion was evaluated. Reddy and Veershetty [4] have come up with the viability analysis of solar parabolic dish collector for Indian conditions. Among the concentrating solar power systems, the parabolic solar dish collector is one of the most efficient energy conversion technologies. Sustainable energy generation can be achieved by the design and implementation of solar parabolic dish power plants. Reddy and Sendhil Kumar [5] numerically studied the combined laminar natural convection and surface radiation heat transfer of parabolic solar dish collector in a modified cavity receiver.

Arulkumaran and Christraj [6] presented experimental platform where steam is generated by non-tracking paraboloid solar dish collector. The design, development and performance characteristics were done in this experiment. Sagade and Shinde [7] evaluated the performance of the solar parabolic dish collector for industrial purpose. From Indian point of view there is a huge possibility for low cost solar water heating systems. The needs of the industrial sector can be fulfilled by this system. El Ouederni et al. [8] explains about the experimental study they had done on parabolic solar collector. The obtained results describe correctly the awaited physical phenomenon. Taylor et al. [9] presented the applicability of nanofluids for various type of solar collectors. The introduction of nanoparticles to conventional working fluids (i.e. nanofluids) can enhance heat transfer and solar collection. The paper explains about a notional design of this type of nanofluid receiver. Sayantan Mukherjee and Somjit Paria [10] presented a brief review to provide an update about the preparation and stabilization methods of nanofluids.

2.1 Inferences from the Review of Literature

From the literature review, it is inferred that no work has been done in order to estimate the characteristics of a parabolic dish collector with nanofluid. An attempt has been made to conduct performance evaluation of dish type solar collector with nanofluid of different volumetric concentrations for various operating conditions.

3 Experimental Studies

3.1 Experimental Set Up

The schematic representation of the experimental set up and its photographic view are shown in Figs. 1 and 2, respectively. The concentrator consists of a dish type frame and a stand. The material used for frame is Aluminium. The stand is made up of mild steel bars and bushes arc provided for proper seating of bearings. Reflectors are one of the important components that are of greater importance for a parabolic dish concentrator. The reflector used in the experimental set up is Aluminium foil which is cheap, durable, long life can cut using latest techniques to required shape and size.

The absorber tube has a diameter of 75 mm and a height of 200 mm. Copper is used because of its high thermal conductivity. The cylinder is placed at the focal point of the concentrator, which has an inlet and an outlet which consists of two thermocouples of K-type for reading the temperature of the water at two different

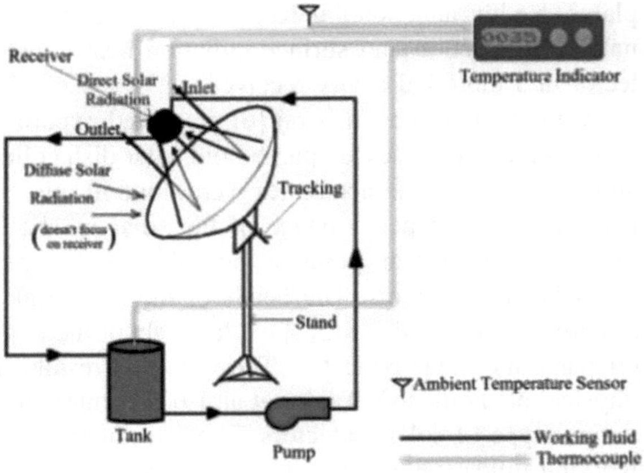

Fig. 1 Schematic of the experimental setup

Fig. 2 Photographic view of the experimental setup

points and a third one is provided for measuring the temperature of the fluid in the tank. The focal length of the collector is 298 mm.

3.2 Experimental Procedure

The absorber is fixed in the focal point of the concentrating collector. Fluid (Water/nanofluid) from the storage tank is allowed to flow through the absorber tube. For all the fluids a constant mass flow rate of 0.058 kg/s and the average radiation of 560 W/m^2 was considered. The following observations are taken for every half-an-hour: (1) Solar insolation. (2) Ambient temperature. (3) Water inlet temperature. (4) Safer outlet temperature. Mechanical tracking mechanism is used so as to tilt the concentrator according to the sun's movement to trap the radiations. The experiment is performed in a setup which has an Aluminium sheet as reflector. A copper absorber cylinder is used. Observations are taken first without Nanofluid that means with water as working fluid and then it is taken with nanofluid of volumetric concentration 0.01 and 0.02% one after another.

3.3 Nanofluid Preparation Process

The first step for experimental studies of nanofluid is the preparation of nanofluid. The two primary methods of nanofluid preparation are single-step and two-step process. In two-step preparation process base fluids are mixed with commercially available nanopowders. These nanopowders are obtained by different physical, chemical and mechanical routes such as grinding, milling and sol–gel and vapour phase methods. This is the mostly used method.

The device used to stir the nanopowders with host fluids is the ultrasonic vibrator as shown in Figs. 3 and 4. Frequent use of ultra-sonication or stirring is required to reduce particle agglomeration. Two-step method can be used to produce alumina nanofluids. Two-step process is the most suitable process to prepare nanofluids which contains oxide nanoparticles other than those containing metallic nanoparticles. The major inherently related issue to this operation is stability because these powders easily agglomerate due to the strong van der Walls forces among nanoparticles.

Fig. 3 Photographic view of the sonicator

Fig. 4 Sonication process

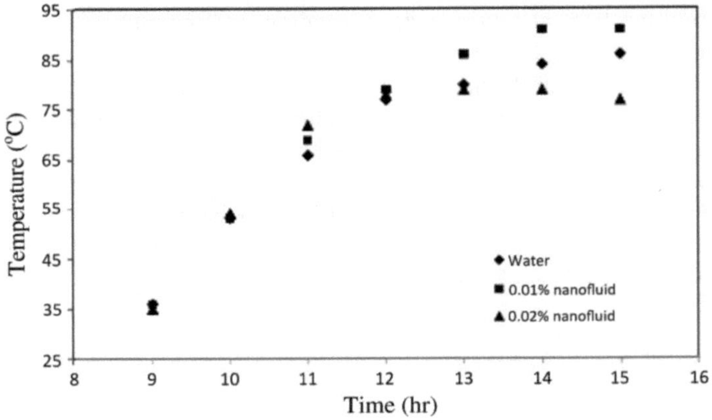

Fig. 5 Variation of outlet temperature with time

4 Results and Discussion

The variation of outlet temperature with respect to time is shown in Fig. 5. It shows that there is a considerable amount of increase in the absorber tube outlet temperature when nanofluid (Alumina-water) is taken as working fluid. Since nanofluid is introduced there is an increase in temperature of about 5–6 °C for 0.01% volume concentration.

From the above graph, it is inferred that when volume concentration increases outlet temperature also increases with respect to time. Since the amount of nanoparticles in the nanofluid increased which leads to increase in the thermal conductivity of nanofluid results in higher outlet temperature when compared to water. The variation of tank temperature with respect to time is shown in Fig. 6

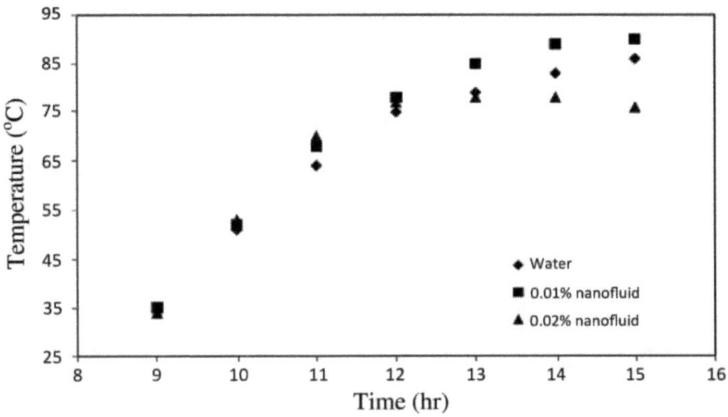

Fig. 6 Variation of tank temperature with time

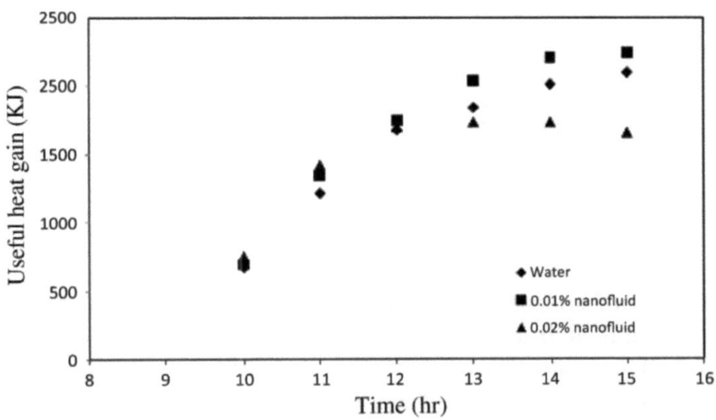

Fig. 7 Variation of useful heat gain with time

which indicates that there is a considerable amount of increase in the tank temperature when nanofluid is used instead of using water as working fluid. Since nanofluid is introduced there is an increase in temperature of 6 °C for 0.01% volume concentration. The useful heat gain difference with respect to time is shown in Fig. 7. It shows that there is a temperature difference due to nanofluid of 0.01% volume concentration.

From the graph, it is clear that when volume concentration increases useful heat gain also increases with respect to time. Since the amount of nanoparticles in the nanofluid increased which increases the thermal conductivity of nanofluid and therefore, considerable amount of increase in heat gain is obtained. The transmittance of the absorber material affects the thermal efficiency of the system.

The measured and estimated parameters of uncertainities are given in Table 1.

Table 1 Details of uncertainties

S. No.	Measured/estimated parameter	Uncertainty
1	Temperature	±0.5 °C
2	Mass flow rate	±0.20%
3	Radiation	±0.50%
4	Useful heat gain	±3.26%
5	Collection efficiency	±3.30%

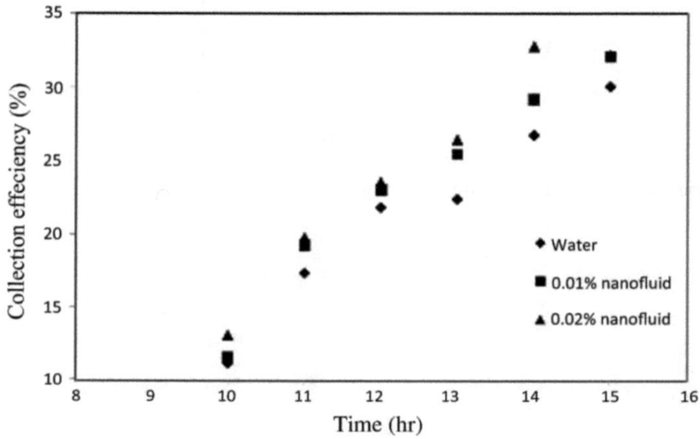

Fig. 8 Variation of collection efficiency with time

The variation of collection efficiency with respect to time is shown in Fig. 8 which shows that there is a considerable amount of increase in collection efficiency when nanofluid is applied instead of water as working fluid. Since nanofluid is introduced there is an increase collection efficiency of about 2–3% for 0–0.02% volume concentration. It is inferred from the experimental studies that when volume concentration increases collection efficiency also increases with respect to time. Since the amount of nanoparticles in the nanofluid increased which increases the thermal conductivity of nanofluid and so considerable amount of increase in collection efficiency is obtained. The amount of solar energy that is concentrated on the receiver of the collector and before the heat is transferred to the liquid that flows along the receiver. Solar energy is drawn by the nanoparticles and transferred to the liquid. The thermal conductivity of nanofluid is better than that of water. Therefore, use of nanofluid gives better collection efficiency.

5 Conclusion

An experimental investigation on the of a parabolic dish type solar collector with nanofluid is done in this study. System's performance has been compared with and without nanoparticles (aluminium oxide). It is observed that, for a volume concentration of 0.01%, the tank temperature reached a maximum of 90 °C with collection efficiency of 32.08%. When the volume concentration is increased to 0.02% a maximum collection efficiency of 32.8% as compared to the 30.07% achieved without nanofluid, has been obtained. It is inferred from Figs. 5, 6, 7 and 8 that when the nanoparticle is introduced with 0.01% concentration, the output parameters such as temperature, useful heat gain and collection efficiency of the system increases compared to that of water. However, when the volume concentration of nanoparticle is increased from 0.01 to 0.02%, the output parameters do not increase due to good stability exhibited by nanoparticles at 0.01% optimum concentration.

References

1. L. Crespo, Z. Dobrotkova, C. Philibert, C. Richter, *Concentrating Solar Power*. International Renewable Energy Agency, pp. 1–48 (2012)
2. L. Li, S. Dubowsky, A new design approach for solar concentrating parabolic dish based on optimized flexible petals. Mech. Mach. Theory, 1536–1548 (2011)
3. S.-Y. Wua, L. Xiao, Y. Cao, Y.-R. Li, A parabolic dish/AMTEC solar thermal power system and its performance evaluation. Appl. Energy 452–462 (2010)
4. K.S. Reddy, G. Veershetty, Viability analysis of solar parabolic dish stand-alone power plant for Indian conditions. Appl. Energy, 908–922 (2013)
5. K.S. Reddy, N. Sendhil Kumar, Combined laminar natural convection and surface radiation heat transfer in a modified cavity receiver of solar parabolic dish. Int. J. Therm. Sci., 1647–1657 (2008)
6. M. Arulkumaran, W. Christraj, Experimental analysis of non tracking solar parabolic dish concentrating system for steam generation. Eng. J. 1–6 (2012)
7. A. Sagade, N. Shinde, Performance evaluation of parabolic dish type solar collector for industrial heating application. Int. J. Energy Technol. Policy, 80–93 (2012)
8. A.R. El Ouederni, M. Ben Salah, F. Askri, M. Ben Nasrallah, F. Aloui, Experimental study of solar concentrator. Revue des Energies Renouvelables, 395–404 (2009)
9. R.A. Taylor, P.E. Phelan, Applicability of nanofluids in high flux solar collectors. J. Renew. Sustain. Energy, 1–8 (2010)
10. S. Mukherjee, S. Paria, Preparation and stability of nanofluids—a review. IOSR J. Mech. Civil Eng., 63–69 (2013)

Part IV
Coating

Macro-study on Hard Anodised Aluminium Oxide (HAAO) Coated Solar Receivers

Ashutosh Arora, Chandra Mohan Arora, Pulak Mohan Pandey, Ashok Kumar Dargar, Suparna Mukhopadhyay and Tapan Kumar Ray

Abstract This paper is discussing development of Hard Anodized Aluminium Oxide (HAAO) coated Solar Receivers with varying degree of surface roughness and finishing of HAAO coatings for comparative analysis. It is found that for lower DNI, roughness Ra < 1.8 μm does not adversely impact much on HAAO coating thermal behaviour, but the dark black polished coating may perform better then greyish unpolished surface, therefore high thickness of HAAO coating is desirable >150 μm using 1000 A rating rectifier (which is limited to 50–60 μm in presented experiment with 400 A rectifier) for further textural developments. Also, their stability at maximum temperature of the focal area of solar concentrators is ~660 °C. HAAO coatings perform better under toughened glass and expected to improve further under vacuum and submerged under water.

Keywords Coating · Anodized alumina · Receivers · Absorption · Emission

A. Arora (✉)
NTPC Limited, New Delhi, India
e-mail: ashutosh.arora@iitdalumni.com

A. Arora
Sir Padampat Singhania University, 248, Sector-4 Hiran Magri, Chanakyapuri, Udaipur, Rajasthan, India

C.M. Arora
Electrical Engineering Department, MNIT Jaipur, Jaipur, Rajasthan, India

P.M. Pandey
Department of Mechanical Engineering, IIT Delhi, New Delhi, India

A.K. Dargar
Faculty of Mechanical Engineering Department, Sir Padampat Singhania University, Udaipur, Rajasthan, India

S. Mukhopadhyay · T.K. Ray
Faculty of PMI, NTPC Limited, Noida, Uttar Pradesh, India

L. Chandra and A. Dixit (eds.), *Concentrated Solar Thermal Energy Technologies*, Springer Proceedings in Energy, https://doi.org/10.1007/978-981-10-4576-9_12

1 Introduction

India is aiming 100 GW of Solar Power by 2022, mainly through PV whereas in future, Solar thermal also being aimed for various applications to get an overall efficiency. This would enable India for partial fulfillment of International Energy Agency (IEA)'s vision 2050 to avoid 2.3 Gt of CO_2 emissions per year.

India has vast solar energy potential of 5000 trillion kWh per year energy with most parts receiving 4–7 kWh/m^2 per day. Due to intrinsic variability nature of solar irradiations, almost all solar thermal and PV systems need selective coatings. Kennedy [1] has reviewed mid to high temperature solar-selective absorber materials in the technical report. The efficiency improvement and cost minimization is to be achieved by increasing the operating temperature of the solar focal area, for which more efficient selective coatings are needed that have both high solar absorptance and low thermal emittance at 500 °C or even higher temperature operation. The coatings need to be stable in air in case the vacuum is not maintained. For efficient photo thermal conversion solar absorber surfaces must have high solar absorptance (α) and a lower thermal emittance (ε) at the operating temperature. A low reflectance ($\rho \approx 0$) at cutoff wavelengths (λ) and a high reflectance ($\rho \approx 1$) at λ characterize spectrally selective coating (SSC). The cutoff may be higher or lower as it is dependent on the temperature [1].

The operational temperature ranges of these materials for solar applications are categorized as low temperature ($T < 100$ °C), mid-temperature (100 °C < $T < 400$ °C) and high temperature ($T > 400$ °C). Parabolic dish concentrators operate at high- and mid-temperatures. The commercial standard of ideal spectrally selective surface would be low cost and easy to manufacture, chemically and thermally stable in air at elevated temperatures ($T \geq 500$ °C) and have solar absorptance ≥ 0.95 and a thermal emittance ≤ 0.2 at 500 °C. Many such coatings have been developed in various researches cited in the literature [2], but how much proven are the cited results, in field application, remain unclear at the user end, therefore advised by Kennedy [1] to investigate further for CSP applications.

Kennedy et al. [1–3] has reviewed the selective solar coating materials in depth. AAO Coatings have been mentioned as spectrally selective by Kennedy [1]. Various designs of the solar radiation receivers and SSCs have been discussed in the Solar Publications and Reports but India is still awaiting 'Make in India' drive for the same to get indigenous resourcing of SSCs.

Sergeant et al. [4] has shown high performance solar-selective absorbers using coated sub-wavelength gratings. anodizing.org, almecosolar.com, alanod.com, etc. sites have elaborated various developments in AAO coatings and as SSCs as well.

2 HAAO Coating Receivers: Development

In the Kennedy's [1] list of mid-temperature selective coating, Anodized Aluminium Oxide (AAO) coating offers high absorptance up to 0.85–0.97 and also characterizes optimally for selective coating as it has low emittance up to 0.08–0.21, operating in open air mid-temperatures 300–400 °C, hence was focused to develop in present work. Higher temperature AAO coatings are possible with RF Sputtering process when alloyed with Ni, Co, Mo, W and Pt.

If the coating development/application costs too much higher than the return of efficiency enhancement expected, then it may not be commercially viable.

Variety of surface textures were created on the HAAO sample pieces, but the variation in Thermal behaviour in varying Surface Roughness of samples are not so remarkable ($\sim$1–2 °C on single sun at DNI $\sim$ 500 W/m^2) therefore no specific surface roughness Ra < 1.8 can be suggested as optimal. There is a possibility of inferior performance of the higher roughness as the visible darkness of the AAO surface reduces from black to grey. Although smoother finished AAO coated surface shines more than the rough surface, hence reflect more of the incident solar radiation, but still it performs better than the rough AAO coated surface due to more blackish surface as compared to greyish surface of rough unpolished AAO coated surface. Therefore, it is recommended to apply black AAO coatings even if roughness is to be sacrificed to achieve the same, and best if thicker coating $\sim$150 μm or even higher as much as possible is made.

3 Comparative Macro-study of HAAO Coatings

Six sample pieces developed for comparative performance based on surface roughness. First sample is of raw Aluminium rolled sheet drawn in saucer shaper without any surface modification. Other five samples are modified at outer and inner surface by emery paper #80 (smoothest available), #120 (course grade hard) and #180 (roughest available) with different combinations as shown in Table 1. That means first sample is just in raw form, the second's outer face made rough with #80 emery paper and inner with #180 emery paper and so on. The darkness as judged visually is marked dark or low with gloss or matt surfaces.

Process parameters for coating development are shown in Table 2. After roughness, all the six samples were coated with hard anodized coating and compared for surface temperatures attained in exposing outer surface to the Sunlight as shown in Table 3 for Macro-Study.

From Table 3, it is evident that the maximum temperature is mostly attained by sample #2 and 3; both are smoother in outer surface as compared to the other samples, although there is no remarkable difference in the temperatures attained by all the samples, it implies that the maximum temperature can be attained by some degree of roughness on the surface. The raw Aluminium rolled sheet when drawn in

Table 1 AAO coated samples of varying surface textures (roughness)

Sample #	1	2	3	4	5	6	Weight
Outer emery #	X	80	80	120	180	180	425 g
Inner emery #	X	180	80	180	80	180	(Each)
R_a outer face	0.5207	0.5074	0.8359	0.751	1.4598	1.1541	(Telisurf)
R_a inner face	1.1097	4.3233	1.6022	2.243	3.785	3.3657	(Telisurf)
Darkness	Dark grey	Dark matt	Dark matt	Low	Low	Low	Visual

Table 2 AAO coated samples development anodizing process parameters

Rectifier current	400	A
Rectifier voltage	80	VDC
Oxalic acid	10	%
Sulphuric acid	15	%
Time retained	3	Hours approx (for 55–60 μm coating)
Bath temperature maintained	−2	°C
Aluminium sheet thickness	3.15	mm

Table 3 Surface temperatures of samples under direct Sun on a sunny day

Time	1	2	3	4	5	6	Time	1	2	3	4	5	6
08:30	43.9	45.8	45	44.5	44.2	46	16:47	54.3	56.6	56.5	55.6	55.5	55.4
09:00	49.3	52.1	49.7	48.3	48.5	48	17:30	43.3	42.7	42.2	42.2	42.9	42.2
09:36	56.7	62	59.8	58.8	60.7	58.7							
10:00	61.8	67.9	65.9	65.7	66.5	64	13:45	83.7	84.5	82	83.1	83.6	83.1
10:30	58	65.5	62.6	62.7	62.3	61	14:30	81	83.5	80.7	80.2	81.9	81.4
11:00	59.5	68.9	66.8	67.1	68	67.1	15:15	68.1	69.4	69.3	70.3	71.4	69.7
11:30	59.8	70.4	68.1	67.4	68	67.4	20:15	37.4	36.6	36.6	37.4	37.3	37.7
12:00	59.9	69.1	67.4	67.1	66.4	66.5							
12:30	59.8	70.3	68.8	67.4	68	68.1	09:45	69.6	70.9	70	70	70.9	69.7
13:23	65.7	70.5	72.5	70.1	68.9	70.3	12:20	69.6	70.6	70	71.7	73.2	71.9
14:45	61	63.1	64.3	62.9	61.1	60.4	13:35	63	63.4	63	62.5	63.4	63
15:15	58.4	65.2	66	64.2	63	64.1	14:45	76.2	77.4	76.8	76.1	77.2	77
16:00	50.7	54.2	55.1	54.3	53.3	52.4	20:45	41.7	41.7	41.7	41.7	41.7	41.7

the die, becomes very glossy (high $\rho \gg 0$) hence there in lesser solar energy collection on the surface, even when the darkness of the coated film on the surface is maximum, i.e. higher absorptance. The sample #2 and #3 are nearly darker to sample #1, but due to the surface roughness, these have become matt instead of remaining glossy hence lesser reflectivity is possible in these samples which retain more solar radiation hence better surface temperature. With this reason, sample #4, 5 and 6 should have gained higher temperatures as they are more rougher than #1, 2 and 3 but it is not happening so because of the fact that darkness of the film is

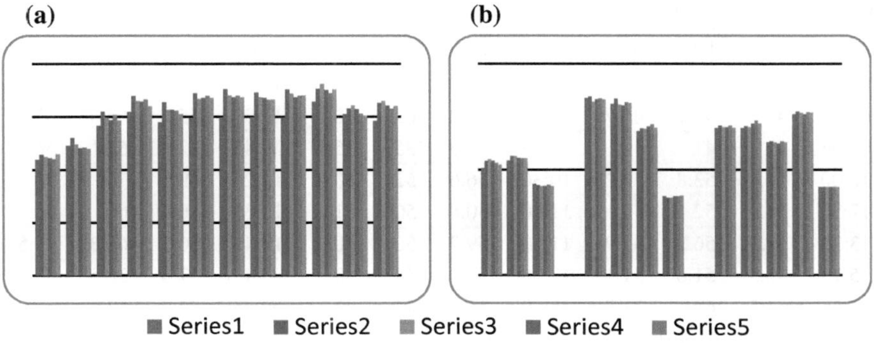

Fig. 1 **a**, **b** Comparative performance of HAAO receiver samples 1–6 at different solar radiations in a day (*X*-axis time in hours and *Y*-axis temperature in °C)

getting reduced in these samples due to lesser penetration of the coating under the surface. This can be improved with the use of 1000 A rating rectifier, for developing thicker coating of AAO.

On other days, sample #5 and 6 also attained higher temperature equally comparable with sample #2; during non-peak DNI hours, therefore roughness has impact on absorptance of the AAO coating at lower DNIs but at higher DNI and higher concentration ratio, it is not necessary to impart high surface roughness on AAO coating. Emery paper #80 would be enough to improve the solar absorptance. The same observation can be better seen in Fig. 1.

Since the rectifier's highest current rating ~400 A have been used, therefore more thicker coating above 60 μm is possible only at 1000 A rating Rectifier, which was not available during this project schedule, therefore the same can be taken up in future developmental work.

After AAO coating, polishing operation is possible on AAO coating, which reduces surface roughness and improves darkness. Improving darkness would give better solar radiation collection efficiency but the reduced roughness would also increase the emittance hence net gain in solar radiation can be optimized by striking balance in roughness, coating thickness and polishing operations.

To test the effect of polishing, three other samples of varying polish exposed to Sunlight, observations recorded in Table 4.

Although in Table 4 and Fig. 2, mixed trend is being observed on different days. In graphical representation, it is observed that no much appreciable difference in polished versus unpolished/semi polished surface existing, therefore it can be concluded that choice of polishing the AAO surface as finished product may be dropped to reduce the cost of the process as well to reduce the reflection of the incident radiation.

To compare the effect of anodizing along with glass casing, four samples chosen, First of MS plate, Second of Plain Aluminium, Third of Anodized, outer surface polished and inner surface unpolished without surface roughness development kept

Table 4 Effect of polishing after anodizing process on AAO surface

Time	P	UP	SP	Time	P	UP	SP	Time	P	UP	SP
08:30	43.4	41.2	40.7	14:30	67.2	68	69	16:47	49.9	49.9	49.9
09:00	43	41.7	40.7	15:15	63.9	62	60.1	17:30	55	55.4	55.4
09:36	41	41	48	20:15	35.3	35	35.2	09:45	62.9	53	56
10:00	56.4	53.8	54.2	10:30	56.6	52	50.6	12:20	69.5	70	70
12:30	54.9	52.1	50	11:00	50.8	50	51.1	13:35	56.6	60	58
13:23	59.5	56.2	55.9	11:30	59.7	52	50.6	14:45	66.7	66.3	69.5
15:15	55.5	51.6	51.1	12:00	57.4	54	50.3	20:45	37.4	36.5	37.2
13:45	71.5	69.7	71.8	16:00	51	47	47.7	09:45	62.9	53	56

P Polished; *UP* unpolished; *SP* semi polished

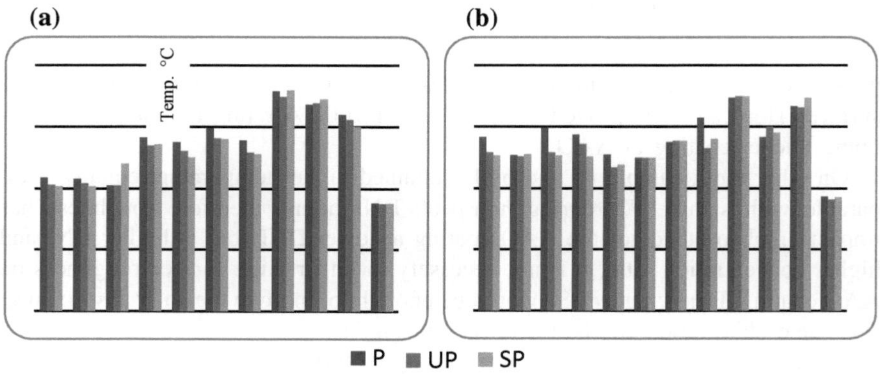

Fig. 2 **a**, **b** AAO coating performance—different polishing surfaces

in open direct contact with atmosphere and last sample also same as third one but kept inside a glass casing. All were kept under sunlight and observations recorded as in Table 5.

For analysis, in Fig. 3 of Table 5 data shows a general trend of increasing surface temperature from Aluminium to MS plate to open AAO to encased AAO surfaces. This implies that for Solar thermal applications it is better to replace MS plate receiver with AAO surface, preferably under a Glass casing.

If complete casing of the AAO surface inside glass is not feasible at site, possibility can be to only cover up the face of AAO coating with Glass Sheet, which should be toughened to sustain higher temperatures of the AAO surface at the focal area of solar radiation receiver or concentrators.

The samples used to experiment are shown in Fig. 4. All temperatures recorded with infrared thermometer of range −30 to +500 °C with an accuracy of ±3 °C. It is desirable to use a higher accuracy thermometer for above experiments to arrive on more accurate results; otherwise these experiments may also be done on higher concentration ratios for wider difference in temperature range of the samples. With

Table 5 Effect of anodized coating on surface with casing inside glass

Time	Al	MS	AAOA	AAOG	Time	Al	MS	AAOA	AAOG
8:30	28.5	35.5	42.8		16:47	33.2	49.2	48.6	58.5
9:00	16	35.5	43.9		17:30	30.6	55.4	42	48
9:36	37.6	37.6	45.5	63.9	13:45	36	41.5	45.9	74.6
10:00	28.2	46.2	55	72	14:30	41.8	60.9	65.8	75.8
10:30	32.7	41.3	50.5	66	15:15	37.7	51.9	43.4	69.9
11:00	29.7	42	50.5	55.7	20:15	37.1	36	36.1	39.1
11:30	33.5	40.6	54.4	71.9					
12:00	36.6	52.4	52.8	71.5	9:45	52	65	64.5	65
12:30	34.2	40	55.1	73.2	12:20	52	59.3	61.4	66.7
13:23	31.8	55.9	57.2	76.9	13:35	48.8	55	55.3	53.3
15:15	35	56	53	69.5	14:45	51.1	58.6	65.7	72.3
16:00	23.1	50.1	47.5	58.9	20:45	36	32.3	37.6	40.3

Al Aluminium; *MS* mild steel; *AAOA* AAO in air; *AAOG* AAO in glass

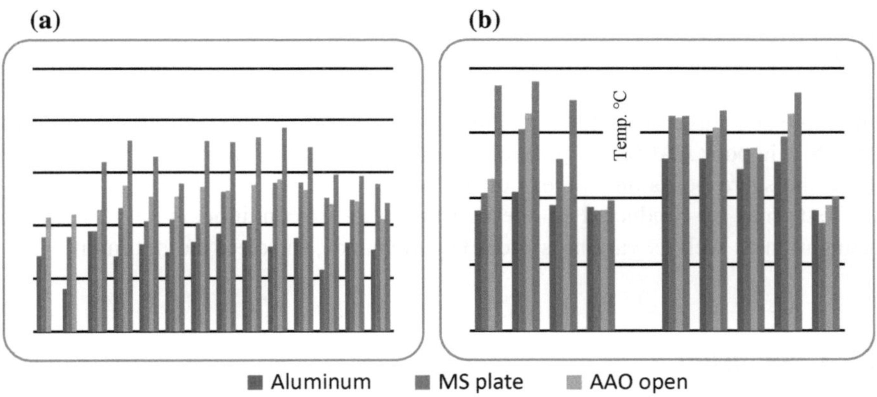

Fig. 3 **a**, **b** Anodized coating on surface with casing inside glass

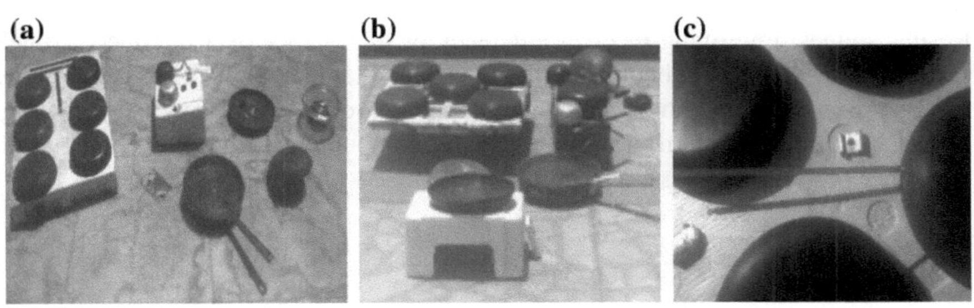

Fig. 4 **a**, **b**, **c** Photographs of samples developed

16 m^2 Scheffler dish, it has been observed that without heat transfer from the coating surfaces, the substrate metal, i.e. Aluminium gets melted instantaneously ~660 °C but HAAO Coatings remains stable.

4 Results and Discussion

Indigenous HAAO coatings developed for low and mid-temperature applications. The AAO coatings should be at least 55–60 μm thick and may be given surface roughness with emery paper #80 for better performance. After anodizing, polishing may be a choice of aesthetics but it may or may not improve performance unless coating is developed thicker ~150 μm for which rectifier rating MCR should be ~1000 A, of commercial grade.

The glass covers can be applied with anti-reflection coating for highest solar absorption together with lower thermal emissivity. The glass should be toughened for sustaining higher temperatures. If the process operational temperature is below 200 °C, then carbon black or powder coating can also be applied for better absorptive properties.

Minimum diameter of Aluminium flat disc needed was 280 mm. Below this, AAO coatings are not enough 55–60 μm thick and hence getting peeled off therefore not stable. Also, it is found that more surface roughness of Ra > 1.8 μm is resulting in poor AAO coating development and its stability as it peels off very soon from the sharp peaks and valleys of the surface.

Tests may be conducted on-site in actual weather conditions with varying heat generation as well as varying system load/demands, to select the optimum.

5 Conclusion

Hard anodized Aluminium coatings may offer a better option for solar thermal applications up to 500 °C, making them selective optically by integrating with anti-reflection coated suitably toughened glass and water submerged absorbers. Results suggest improving receiver efficiency by ~10–15% over the uncoated MS plate surfaces along with better stability at elevated temperatures. The cost of AAO receiver may be justified over the conventional MS receiver with any other coating, in fact in mass production; it may be equally competitive with MS receivers with any other type of coatings.

6 Future Prospects and Suggestions

Thicker HAAO coating with vacuum glass shields may be further experimented with higher solar radiation concentration ratio; to arrive on solutions for high temperature applications. There was limited time period for which the data collected and analyzed. It can be done over a year for a particular location, so that better conclusion for exact specifications of particular type of coating selection can be drawn up. HAAO solar receivers can relive parabolic dish users from the corrosion and erosion troubles being faced with conventional receivers, in addition to enhance the energy efficiency of the solar receivers.

Acknowledgements Authors are thankful to the Administrative/Managerial Authorities and Faculties of IIT Delhi, MNRE-NISE, WRST, Prajapita Brahmakumaris Ishwariya Vishwa Vidhyalaya, IIT Jodhpur, Hindalco, Kiran Cookware Industries, IGSTPP/APCPL NETRA and PMI of NTPC Limited, TERI, BHEL, Thermax Limited, HCL and SPS University Udaipur for extending their Cooperation and Coordination in the study work presented and in continuation further.

References

1. C.E. Kennedy, Review of mid- to high-temperature solar selective absorber materials review of mid- to high-temperature solar selective absorber materials. Contract, July. Retrieved from http://large.stanford.edu/publications/coal/references/troughnet/solarfield/docs/31267.pdf (2002)
2. C.E. Kennedy, Progress to develop an advanced solar-selective coating. Renew. Energy, 4–7. Retrieved from http://large.stanford.edu/publications/coal/references/troughnet/solarfield/docs/36997.pdf (2008)
3. C. Kennedy, H. Price, Development and testing of high-temperature solar selective coatings. Sol. Energy, 322–323. NREL (2005)
4. N.P. Sergeant, M. Agrawal, P. Peumans, High performance solar-selective absorbers using coated sub-wavelength gratings. Opt. Express **18**(6), 5525–5540. OSA. https://dx.doi.org/10.1364/OE.18.005525 (2010)

Websites

5. www.alanod-solar.com; www.anodizing.org; www.india-one.net; www.almecosolar.com
6. https://en.wikipedia.org/wiki/Concentrated_solar_power; www.bksew.org
7. www.mnre.gov.in; www.ntkj.co.jp; www.solarconference.net; www.seci.gov.in
8. www.pfonline.com/articles/new-uses-for-anodized-Aluminium-oxide

Effect of Growth Condition on Mechanical Properties of Zirconium Carbonitride Absorber-Based Spectrally Selective Coatings

B. Usmani, V. Vijay, R. Chhibber and Ambesh Dixit

Abstract Zirconium carbonitride (ZrC–ZrN) absorber layer has been optimized for the maximum absorptivity $\alpha \sim 0.9$ in ZrO_x/ZrC–ZrN/Zr reflector–absorber tandem structures on stainless steel (SS) and aluminum (Al) substrates, using reactive DC/RF magnetron sputtering process and varying reactive nitrogen during the deposition. The mechanical properties such as hardness and Young's modulus values increase from ~19.63 to ~38.53 GPa and ~36.39 to ~58.67 GPa, respectively, with increasing nitrogen flow rate during ZrC–ZrN absorber layer deposition on SS substrate. These values also increase with increasing load up to a critical limit and saturates thereafter with further increase in load values. The films with moderate hardness and Young's modulus exhibit enhanced solar thermal performance ($\alpha \sim 0.9$) against films with lower and higher values of these mechanical properties, suggesting optimal nitrogen concentration for desired solar thermal performance.

Keywords Spectrally selective coatings · Nanoindentation · Young's modulus · Hardness · Absorptance · Emittance

B. Usmani · A. Dixit (✉)
Department of Physics and Center for Solar Energy,
Indian Institute of Technology Jodhpur, Jodhpur 342011, Rajasthan, India
e-mail: ambesh@iitj.ac.in

V. Vijay
Department of Mathematics, Indian Institute of Technology Jodhpur, Jodhpur 342011, Rajasthan, India

R. Chhibber
Department of Mechanical Engineering, Indian Institute of Technology Jodhpur, Jodhpur 342011, Rajasthan, India

L. Chandra and A. Dixit (eds.), *Concentrated Solar Thermal Energy Technologies*, Springer Proceedings in Energy, https://doi.org/10.1007/978-981-10-4576-9_13

1 Introduction

There are continuous efforts to design and develop spectrally selective coating, which may withstand high-temperature and harsh environmental conditions. The structural, thermo-physical, mechanical, and corrosion stabilities are essential for the longevity of solar thermal systems. Recently, zirconium carbonitride based spectrally selective coatings have shown promise because of their high-temperature stability and moderate corrosion resistance [1–3]. In addition, the mechanical stability of these spectrally selective coating is indispensable for the long-term stability of the solar thermal properties and thus the system performance. The mechanical properties such as hardness and Young's modulus of the spectrally selective absorbers are important to optimize for optimal solar thermal performance. Therefore, there is a need to understand the correlation of mechanical properties with synthesis conditions, which may be used to design the optimal coating structures with enhanced solar thermal performance in conjunction with their mechanical stability.

Atomic Force Microscopy (AFM) is used to measure the repulsive and attractive forces between a given sample surface and an AFM tip, for mapping the topography of the material at nanometer resolution. Further, in contact mode, AFM has been employed to measure the mechanical properties such as hardness and Young's modulus of the thin film structures in nanoindentation configuration [4–7]. These measurements rely on the interdependence of the force applied to the tip and the indentation depth created during nanoindentation. The measured force–distance curve is used to calculate the hardness and Young's modulus [8]. In this work, we will discuss the impact of growth conditions on the hardness and Young's modulus of zirconium carbonitride based spectrally selective coatings. The observed mechanical properties are strongly related to the zirconium nitride fraction in ZrC–ZrN matrix of the tandem absorber structures. The studies suggest that optimal nitrogen flow rate is important to achieve the enhanced solar thermal performance, which corresponds to the moderate mechanical properties. The lower and higher values of hardness and Young's modulus lead to the poor solar thermal performance and may not be useful for field applications.

2 Experimental Details

The synthesis of ZrO_x/ZrC–ZrN/Zr spectrally selective reflector–absorber tandem structures has been described by Usmani et al. in detail elsewhere [1]. In brief, ZrO_x/ZrC–ZrN/Zr structures are deposited on stainless steel, SS, and aluminum, Al, substrates using DC/RF magnetron sputtering in three sequential steps. First, DC magnetron sputtering was used to deposit Zr metallic reflector, followed by reactive RF magnetron sputtering for ZrC–ZrN absorber layer. Finally, ZrO_x layer was deposited using reactive DC magnetron sputtering. The details about their structural, microstructural, optical, and thermal properties are discussed in Ref. [1]. Mechanical properties of ZrO_x/ZrC–ZrN/Zr structures on both SS and Al substrates are investigated using nanoindentation technique. The nanoindenter, used for mechanical measurements, is an integrated accessory with atomic force microscopy (scanning probe microscopy (SPM) XE-70, Park) system. AFM was operated in contact mode for these mechanical properties measurements over $5 \times 5\ \mu m^2$ scan areas. The Berkovich indenter, a sapphire cantilever with a diamond tip, has been used for indentation on ZrO_x/ZrC–ZrN/Zr/Substrates top surface. This indenter was initially forced into the ZrO_x/ZrC–ZrN/Zr surface at constant load with the forward (down) speed of about 0.3 μm/s and was kept for 10 s at that depth, followed by unloading at a backward (up) speed of about 0.3 μm/s. Ten such indentations were carried out on each sample at different physical locations, to measure the average hardness and Young's modulus using the generated load–displacement curves. The measured load versus displacement graphs were used to calculate stiffness S (=dP/dh) by estimating the slope of the upper portion of the unloading curve. The calculated stiffness S and projected contact area of the indented tip A were used to estimate the effective Young's modulus of these multilayer structures using the following equation [9]:

$$\frac{1}{E_{\text{eff}}} = \frac{2\beta}{S}\sqrt{\frac{A}{\pi}} = \frac{1 - v_s^2}{E_s} + \frac{1 - v_i^2}{E_i},$$

where E_{eff} is the effective modulii for each indenter/specimen combination, and β is a shape constant of 1.034 for the Berkovich tip [4]. E and v represent the Young's modulus and Poison's ratio of the indenter, i, and the sample, s, respectively. The values, used for Poisson's ratio v and Young's modulus E_i for diamond indenters, are 0.07 and 1141 GPa [10]. The Poisson's ratio for samples used is ~0.3 [11]. Simultaneously, the hardness values are calculated using $H = \frac{P_{\max}}{A}$, where $P_{\max}$ is the maximum load in a load–displacement curve.

3 Results and Discussion

Sputter deposition parameters are important to optimize for the optimal solar thermal response of spectrally selective absorbers in the desired wavelength range. In our previous work, we have discussed the optimization of spectrally selective ZrO_x/ZrC–ZrN/Zr reflector–absorber tandem structures, by varying nitrogen flow rate for the absorber layer during deposition [1, 2]. The maximum absorptivity $\alpha \sim 0.9$ and the minimum emissivity $\varepsilon \sim 0.04$ have been observed for 12.5 sccm nitrogen flow during ZrC–ZrN layer deposition. ZrN phase fraction was increasing with an initial increase in nitrogen flow rate initially and exhibited maxima at 12.5 sccm, followed by a decrease with any further increase for both SS and Al substrates [1]. The ZrN phase fraction has been calculated using X-ray diffraction

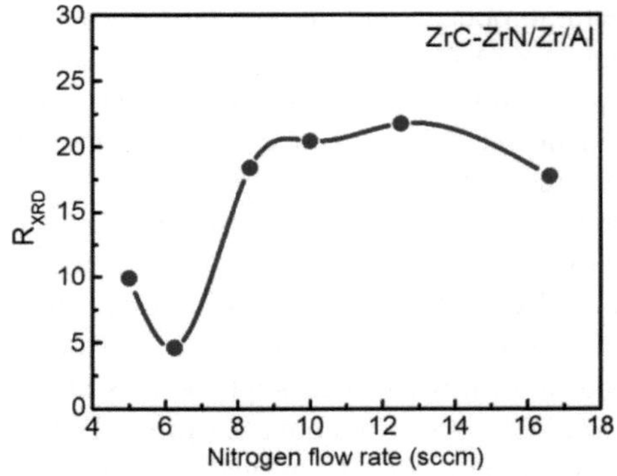

Fig. 1 Zirconium nitride (ZrN) phase fraction in zirconium carbide and zirconium nitride (ZrC–ZrN) matrix as a function of nitrogen flow rate, used during synthesis of the ZrC–ZrN absorber layer

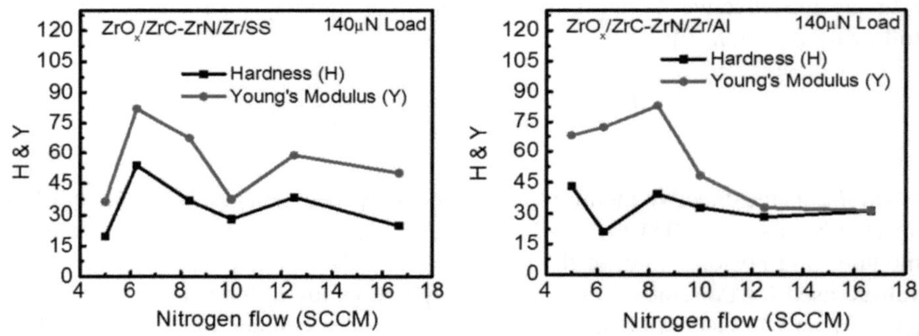

Fig. 2 *Left panel* Hardness (H) and Young's modulus (Y) of ZrO_x/ZrC–ZrN/Zr tandem absorber film on SS substrate and *right panel* on Al substrate

measurements of ZrC–ZrN absorber layer in ZrO_x/ZrC–ZrN/Zr structure on Al substrate. The measured relative ZrN phase fraction in ZrC–ZrN absorber matrix has been shown in Fig. 1, as a function of nitrogen flow rate. This variation in ZrN phase fraction suggests that the maximum is near ~10–12.5 sccm nitrogen flow rate, used for the synthesis of ZrC–ZrN absorber layer.

The mechanical properties are important and assist in understanding the materials strength and durability. The nanoindentation measurements have been carried out to calculate the hardness and Young's modulus of ZrO_x/ZrC–ZrN/Zr structures. The nanoindentation depth ~50 nm is much smaller as compared to the thickness of the entire ZrO_x/ZrC–ZrN/Zr structure ~600 nm. The thickness of ZrO_x/ZrC–ZrN/Zr structure is sufficient to exclude the substrate effect on these measured mechanical properties and thus, the measured values can be considered mostly from the ZrOx/ZrC–ZrN/Zr structures only. The measured values of hardness and Young's modulus are summarized in Fig. 2 for both SS (left panel) and Al (right panel) substrates, as a function of nitrogen flow used for the synthesis of ZrC–ZrN absorber layer. The hardness values show strong variation for ZrO_x/ZrC–ZrN/Zr structures, with lower nitrogen flow rates, used for ZrC–ZrN layer and approached nearly constant value at 10 or higher sccm nitrogen flow rates, for both SS and Al substrates. The hardness values are prone to the film properties such as films' density, microstructure, and surface roughness [12]. Thus, samples with 10 sccm or higher nitrogen flow may be showing less surface roughness with optimal density of ZrC–ZrN composite absorber, causing nearly constant hardness values in contrast to lower nitrogen flow rates, where effective density of ZrC–ZrN may be lower and also surface properties may be poor, leading to large variation in hardness values. The maximum selectivity of ZrO_x/ZrC–ZrN/Zr tandem structure selective absorber was found around 12.5 sccm nitrogen flow rate used for ZrC–ZrN absorber layer [1], suggesting moderate values of hardness may be relatively better for solar thermal performance.

The elastic modulus is a measure of the stiffness of the material and Young's modulus has been used to understand stiffness properties. Young's modulus also exhibits the similar trend like hardness values, which nearly saturates toward higher nitrogen flow rates, in conjunction with initial large fluctuations. The reason of such fluctuation is not clear for samples deposited at lower nitrogen flow rates. Hardness and Young's modulus values of ZrO_2, ZrN, and ZrC single-layer thin film are 13.5–19 GPa, 13.4–23.5 GPa, 27.6 GPa and 197–210 GPa, 166.4–196.3 GPa, 228 GPa, respectively [12–14]. Interestingly ZrO_x/ZrC–ZrN/Zr tandem structures exhibit higher values of hardness as compared to that of ZrO_2, ZrN, and ZrC single-layer thin film hardness values. However, Young's modulus values show contrary behavior with relatively smaller values as compared to ZrO_2, ZrN, and ZrC single-layer thin film values. This is possible as hardness and Young's modulus in the present work is the effective values of multilayer ZrO_x/ZrC–ZrN/Zr structures in contrast to the single layer independently.

Further, we carried out load versus displacement curves at different initial load conditions on the sample with the optimal solar thermal properties $\alpha \sim 0.9$ and $\varepsilon \sim 0.04$, synthesized suing 12.5 sccm nitrogen flow rate to understand the impact of external load on mechanical properties. The measured different loads versus displacement curves are shown in Fig. 3a–e, which are used to measure the hardness and Young's modulus values at these initial load conditions. The respective insets show the microscopic image of the indented area, created during nanoindentation measurements.

These nanoindentation profiles suggest that depth of such indentations increases with increasing load, as noticed with increased diameter and marked in insets for clarity. The measured H and Y values as a function of load are summarized in Fig. 3f for this sample within the elastic limit. These measured values suggest that initially both hardness and Young's modulus values are increasing with load up to ~ 140 μN and reached to nearly constant hardness and Young's modulus values, independent of load conditions.

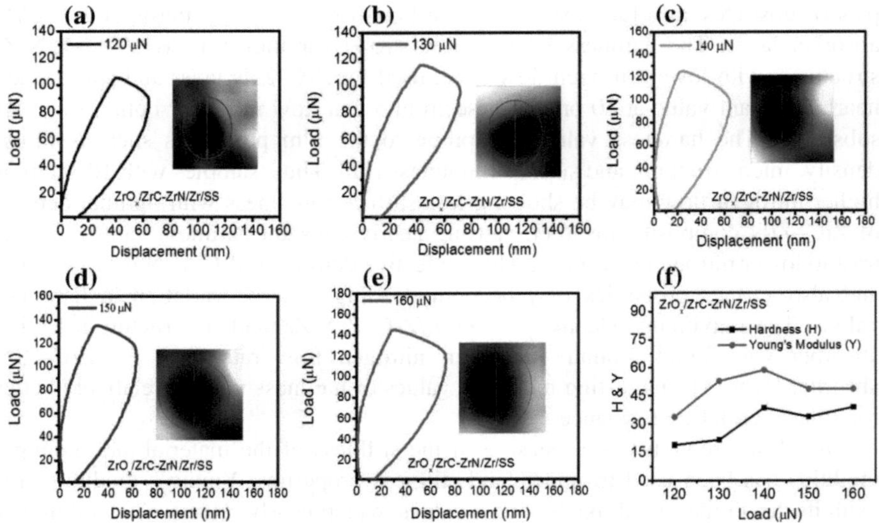

Fig. 3 The load versus displacement curve at difference end loads, **a**–**e** with inset showing the respective AFM indentation profiles and hardness and Young's modulus values versus load, **f** for ZrO_x/ZrC–ZrN/Zr tandem structure with ZrC–ZrN absorber layer synthesizes at 12.5 sccm nitrogen flow

4 Conclusion

ZrO_x/ZrC–ZrN/Zr reflector–absorber tandem spectrally selective surfaces were deposited on stainless steel (SS) and aluminum (Al) substrates, using RF/DC sputtering system with varying nitrogen flow rate during ZrC–ZrN absorber layer deposition. AFM nanoindentation method has been used to understand the mechanical properties of ZrO_x/ZrC–ZrN/Zr structures on both substrates. The hardness and Young's modulus values increase from ~19.63 to ~38.53 GPa and ~36.39 to ~58.67 GPa with increasing nitrogen flow rate of the absorber layer. These mechanical properties also exhibit an increasing trend with varying load up to 140 μN, which becomes constant with any further load increase. These studies suggest that absorber growth conditions have a strong influence on mechanical properties and moderate hardness and Young's modulus values may be important with optimal solar thermal performance.

Acknowledgements Authors gratefully acknowledge the financial assistance, from the Ministry of New and Renewable Energy (MNRE), India through grant 15/40/2010-11/ST, to carry out experimental work.

References

1. B. Usmani, A. Dixit, Spectrally selective response of ZrO_x/ZrC–ZrN/Zr absorber-reflector tandem structures on stainless steel and copper substrates for high temperature solar thermal applications. Sol. Energy **134**, 353–365 (2016)
2. B. Usmani, A. Dixit, Impact of corrosion on microstructure and mechanical properties of ZrO_x/ZrC–ZrN/Zr absorber-reflector tandem solar selective structures. Sol. Energy Mater. Sol. Cells **157**, 733–741 (Accepted, 2016)
3. B. Usmani, V. Vijay, R. Chhibber, L. Chandra, A. Dixit, Zirconium car bide-nitride composite matrix based solar absorber structures on glass and aluminum substrates for solar thermal applications, in ISES Solar World Congress 2015 Proceeding (in press)
4. T.-H. Fang, S.-R. Jian, D.-S. Chuu, Nanomechanical properties of TiC, TiN and TiCN thin films using scanning probe microscopy and nanoindentation. Appl. Surf. Sci. **228**, 365–372 (2004). https://dx.doi.org/10.1016/j.apsusc.2004.01.053
5. S.A. Catledge, J. Borham, Y.K. Vohra, W.R. Lacefield, J.E. Lemons, Nanoindentation hardness and adhesion investigations of vapor deposited nanostructured diamond films. J. Appl. Phys. **91**, 5347 (2002). https://dx.doi.org/10.1063/1.1464233
6. R. Navamathavan, D. Arivuoli, G. Attolini, C. Pelosi, Nanoindentation studies of (111) GaAs/InP epilayers. Appl. Surf. Sci. **180**, 119–125 (2001). https://dx.doi.org/10.1016/S0169-4332(01)00336-1
7. N.R. Moody, W.W. Gerberich, N. Burnham, S.P. Baker, *Fundamentals of Nanoindentation and Nanotribology* (Materials Research Society, Warrendale, PA, 1998)
8. R. Ferencz, J. Sanchez, B. Blumich, W. Herrmann, AFM nanoindenta tion to determine Young's modulus for different EPDM elasto mers. Polym. Test. **31**, 425–432 (2012). https://dx.doi.org/10.1016/j.polymertesting.2012.01.003
9. W.C. Oliver, G.M. Pharr, Measurement of hardness and elastic modulus by instrumented indentation advances in understanding and refinements to the methodology. Mater. Res. Soc. **19**, 3–20 (2004). https://dx.doi.org/10.1557/jmr.2004.19.1.3
10. W.C. Oliver, G.M. Pharr, An improved technique for determining hardness and elastic modulus using load and displacement sensing indentation experiments. J. Mater. Res. (1992). https://dx.doi.org/10.1557/JMR.1992.1564
11. J.F. Shackelford, W. Alexander, Materials Science and Engineering Handbook, 3rd edn. (CRC Press LLC, 2001)
12. C.S. Chen, C.P. Liu, C.Y.A. Tsao, H.G. Yang, Study of mechanical properties of PVD ZrN films, deposited under positive and negative substrate bias conditions. Scr. Mater. **51**, 715–719 (2004). https://dx.doi.org/10.1016/j.scriptamat.2004.06.005
13. D. Craciun, G. Socol, G. Dorcioman, N. Stefan, G. Bourne, V. Craciun, High-quality ZrC, ZrC/ZrN and ZrC/TiN thin films grown by pulsed laser deposition. J. Optoelectron. Adv. Mater. **12**, 461–465 (2010)
14. Z.W. Zhao, W. Lei, X.B. Zhang, B.P. Wang, B.K. Tay, Nanocrystalline zirconium oxide thin films grown under low pulsed dc voltages. J. Phys. D Appl. Phys. **42**, 215408 (2009). https://dx.doi.org/10.1088/0022-3727/42/21/215408

Transition Metal-Based Spectrally Selective Coatings Using In-House Developed Spray System

Ajoy K. Saha, Laltu Chandra and Ambesh Dixit

Abstract The development of transition metal (Fe, Co, Ni, Mn) oxides based thin coatings is reported on stainless steel substrates for solar absorber applications. Absorber layers are synthesized using in-house developed spray system for metal–metal oxide composite coating structures. The optimized combinations of transition metal precursors are sprayed for thin film deposition. The post-spray heat treatment has been carried out to convert the transition metal precursor layers into spectrally selective absorber coatings showing metal–metal oxide composite structures. Three combinations of bi-transition metals Co–Mn, Co–Fe, and Co–Ni precursors are used to synthesize thin film structures. The thickness of these thin films structures is nearly 1 μm. These spectrally selective coatings exhibit high absorptivity ($\alpha \sim 0.9$) in 300–900 nm wavelength range and emissivity ($\varepsilon \sim 0.18$–0.44) in 2.5–25 μm wavelength range. The extension of this work is focused on the development of a large-scale system, capable of fabricating spectrally selective coating on desired structures and optimization of annealing conditions, leading to the optimum solar thermal performance with high absorptivity ($\alpha > 0.95$) and low emissivity ($\varepsilon < 0.1$) in the desired wavelength ranges.

Keywords Solar · Energy · Coating · Spray · Metal · Oxides

A.K. Saha
Center for Solar Energy, Indian Institute of Technology Jodhpur,
Old Residency Road, Ratanada, Jodhpur 342011, Rajasthan, India

L. Chandra
Department of Mechanical Engineering and Center for Solar Energy,
Indian Institute of Technology Jodhpur, Jodhpur 342011, Rajasthan, India

A. Dixit (✉)
Department of Physics and Center for Solar Energy, Indian Institute of Technology Jodhpur,
Jodhpur 342011, Rajasthan, India
e-mail: ambesh@iitj.ac.in

L. Chandra and A. Dixit (eds.), *Concentrated Solar Thermal Energy Technologies*,
Springer Proceedings in Energy, https://doi.org/10.1007/978-981-10-4576-9_14

1 Introduction

The utilization of abundant renewable solar energy is important to reduce the reliance on rapidly depleting fossil fuels. Concentrated solar thermal (CST) technologies are gaining attention, where solar energy is converted into thermal energy for possible applications including electricity generation [1]. CST technologies rely on the design and development of cost-effective solar absorber coatings. An efficient solar absorber coating should exhibit high absorptivity ($\alpha > 0.9$) and low emissivity (<0.1) characteristics over the desired wavelength and operating temperature range [1]. Several techniques are available, including spray, sputtering, electrochemical, and chemical vapor deposition (CVD) [2] for fabricating spectrally selective surfaces. Although physical and chemical vapor deposition techniques have been exploited for solar absorber films, their application is limited to small planar substrates. However, CST technologies require coatings to be developed for larger surfaces like pipes, curved surfaces and specific physical and chemical vapor deposition systems have to be developed for such real applications. In such cases, wet chemical techniques, such as spray and electrochemical techniques, may provide an alternative to fabricating such layered structures. Among the wet chemical techniques, sol-gel [3–5] based dip coating and electrochemical [6–8] routes have been investigated widely for spectrally selective coating applications. However, little literature is available on the spray technique [9, 10], especially for solar selective coatings. The spray pyrolysis technique has been used to fabricate black nickel [9] coatings on Al and galvanized iron substrates. Uma et al. [10] synthesized cobalt oxide-iron oxide coatings using similar techniques. However, considering the high flexibility and cost effectiveness of spray technique for solar absorber coating fabrication, the spray route needs to be designed and optimized extensively, for small and large surface coating applications. Here, we discuss the design and development of a semi-automated spray system, as explained schematically in Fig. 1, which can be used to fabricate the solar absorber coatings on any type of surfaces such as curved surfaces, tubes, pipes, and planar substrates. The system provides length, breadth, and height variability to optimize the quality of the thin films. In addition, the gas flow and precursors flow rates can be used as the control parameters to further improve the film quality and materials' composition for desired structural, optical, and solar thermal properties. Thus, the intended system may provide complete flexibility to engineer the design parameters for desired solar thermal properties and geometrical scalability for coatings on any desired surface. The initial work has focused to demonstrate the feasibility of this process on a smaller scale, where precursor solutions for coatings were sprayed on stainless steel (SS) substrates, 3 cm × 2 cm sizes, using a manually operated spray gun and sprayed structures were subjected to the heat treatment for creating the metal–metal oxide composite matrix for desired solar thermal properties.

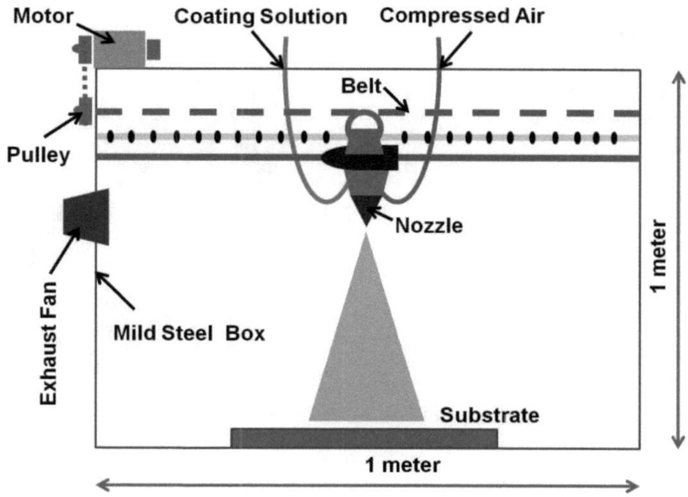

Fig. 1 Schematic of semi-automatic spray coating setup, with legends explaining the respective components used in the system's development

2 Experiment

The spray coating setup consists of a 1 m × 1 m × 1 m stainless steel structural framework. The sides are covered by thin light MS steel plates for protecting the spillage of material during the spray process. This setup includes a motor (Nex Robotics, 170RPM 37DL), driven by L298 Motor Driver Kit and controlled by a microcontroller (Arduino Mega 2560 R3), belt, and pulley (Robokits, GT2 timing pulley 20 teeth, 6 mm width, 5 mm bore diameter) system, which is utilized for to and fro motion of the liquid spraying nozzle along the entire X-axis, as explained in Fig. 1. The system consists of (i) precursor line and (ii) gas line, as marked in Fig. 1. The flow rates of both, precursor and gas can be controlled independently to regulate the end flow rate across the nozzle and thus the film thickness during spray deposition. The base of the spray system has been used for holding substrates without and with a controlled heating assembly for desired processing temperature. The coating precursor solution is sprayed using compressed air, where the distance between substrate and spray nozzle can be adjusted for optimal spraying conditions. In another attempt, some samples were prepared by spraying solutions using a spray gun to test and optimize the precursors spraying conditions. The spray solution was prepared by dissolving transition metal (Ni, Co, Mn and Fe) salts in ethyl alcohol.

Ethanolamine was used as a chelating agent, while polyethylene glycol was added to obtain the optimum viscosity required for spraying the prepared solution. Stainless steel (SS) substrates were cut in sizes of 3 cm × 2 cm from larger plates. They were ground with 2000 grade grinding paper followed by washing with lime and salt. Subsequently, to remove oily contaminants, the ground substrates were

Fig. 2 Optical photographs of spray coated thin films on SS substrates, using transition metals (Fe, Ni, Co and Mn) in ethyl alcohol, heat-treated at 100 °C for 1 h

treated with trichloroethylene at 80 °C for 5 min followed by acetone under similar conditions. The cleaned substrates were then dried by dabbing with disposable tissues. The substrates, thus prepared were mounted on a big steel plate by pasting them with paper tapes for deposition of desired precursors' thin films at room temperatures. The setup was arranged vertically inside a fume hood and then sprayed with the precursor solution using the spray gun. For good quality coating, the optimum spray distance was fixed at about 1 m from the substrates. The spraying was continued to and fro from one end to the other for several times, to achieve the desired thickness of the coated thin films. The coating thickness variation needs to be strictly monitored for optimum coating performance during the spray process using different control parameters such as nozzle to substrate distance, stepping speed, and controlled flow rates of precursor and gas. The optical photographs of the room temperature sprayed thin films are shown in Fig. 2. The structural and microstructural property evaluations of the synthesized materials were carried out using X-ray diffractometer (XRD) and scanning electron microscopic (SEM) measurements. The optical properties were investigated using UV–Vis and Fourier transform infrared (FTIR) spectrophotometers. The obtained measurements were used to estimate the room temperature solar thermal response of these film structures.

3 Results and Discussions

The transition metal (Fe, Ni, Co, Mn) salts in ethyl alcohol were used as precursors and sprayed on these cleaned stainless steel substrates to fabricate CoMn, CoFe, and CoNi thin film coatings. The different combination of bi-transition metal precursors produced different colored coatings.

The color of the as-prepared films produced using Co and Ni precursor salts was reddish, while those prepared from Co and Fe resulted in the greenish color. The color of the films containing Co and Mn was pink, as shown in Fig. 2. These as-prepared coatings were dried in air at room temperature for about an hour, followed by air drying in an oven at 100 °C for 1 h. The coated films retained their original color under such heating conditions up to 100 °C. These thin films spread uniformly over the substrate and exhibited good adhesive properties. These air dried coatings were further heated at 250 and 400 °C for 1 h each in open ambient conditions. The pristine color of these different colored coatings changed into the dark black after 250 °C and higher temperature treatment. This color change is due to the formation of transition metal oxides at higher temperatures, where band gap of transition metal oxide composites gave rise to the absorption and thus the dark black color of these annealed thin films. In contrast, the as-prepared thin films are colored because of intrinsic d-d electronic transitions in different transition metals, corresponding to different wavelengths. The structural evolution of these films has been investigated using XRD measurements with 2θ from 30° to 80° and the results are plotted in Fig. 3. The XRD spectrum of bare stainless steel substrate was also recorded and plotted in the bottom panel of Fig. 3, for comparison, and to distinguish the developed transition metal oxide structures on the SS substrate.

The XRD spectra of bare SS substrates exhibit austenite and ferrite phases, as indicated in the bottom panel of Fig. 3. The fabricated structures, annealed at different temperatures do not show any additional XRD peaks indicating that the transition metal oxide coatings are amorphous in nature. The surface microstructure and chemical compositions of these coatings were investigated using scanning electron microscopy and EDX measurements for heat-treated structures at 100, 250 and 400 °C and micrographs are shown in Fig. 4.

The bare SS substrate, shown in Fig. 4a suggests large cracks developed during the cleaning of the substrate with coarse abrasive paper. The surface micrographs of coatings heated at different temperatures are shown in Fig. 4b–d. The surface SEM

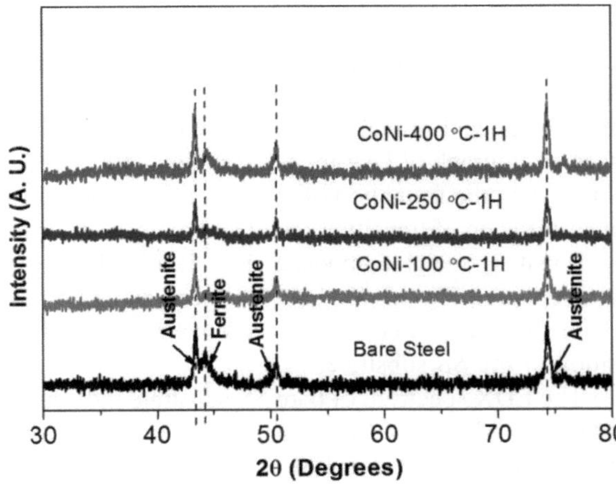

Fig. 3 XRD spectra for CoNi bi-transition metal coatings on SS substrate after heating at 100, 250, and 400 °C for 1 h with bare SS substrate

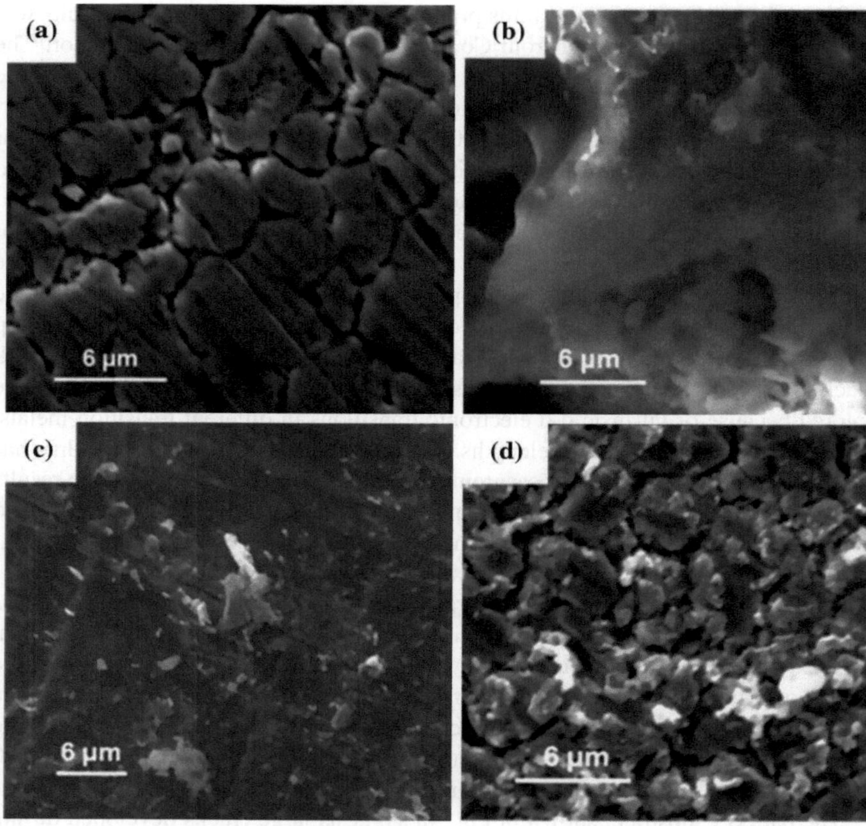

Fig. 4 Scanning electron microscopic images of, **a** bare steel substrate, coatings containing Co and Mn after heating at **b** 100 °C, **c** 250 °C and **d** 400 °C for 1 h

image of 100 °C heat-treated sample, Fig. 4b, is floppy in nature indicating the presence of unburnt organic materials present in the spray precursor solutions. The surface of heat-treated sample at 250 °C, Fig. 4c, shows coating on SS along with the cracks, and granular coverage with larger grain boundaries. In contrast, a surface micrograph of heat-treated sample at 400 °C, Fig. 4d, shows a smooth surface, suggesting the formation of uniform film across the substrate. This is because of the interdiffusion of transition metal and oxygen to form uniform transition metal oxide—transition metal composite thin film uniformly across the substrate at higher temperature. The EDX elemental analysis results are summarized in Table 1 for these films, annealed at various temperatures, together with the chemical composition of the SS substrate.

The EDX spectrum of bare SS substrate indicates the presence of ~74 at% Fe, 9 at% Mn and 17 at% Cr elements. The thin films prepared using Co and Mn metal precursors and heated at 100 °C exhibit ~6.4 at% Co and 1.2 at% Mn respectively,

Table 1 EDX analysis of films containing Co and Mn after annealing at **a** 100, **b** 250 and **c** 400 °C, relative to the bare stainless steel substrate

Coatings	SS with coating (at%)			Bare SS substrate	
Element	100 °C	250 °C	400 °C	Element	at%
C	57.4	–	–	C	–
O	29.1	20.1	38.0	O	–
Cr	1.2	12.4	8.4	Cr	17.1
Mn	1.2	7.4	7.0	Mn	8.9
Fe	4.1	56.0	38.4	Fe	74.1
Co	6.4	4.2	8.3	Co	–

with 57.4 at% C and ~29.1 at% oxygen, confirming the presence of precursor unburnt organic material in the sample. The SEM micrograph of the sample above also confirms the same, where floppy surface morphology has been observed because of unburnt organic contents in the coating. The presence of oxygen does not indicate the conversion of Co and Mn into transition metal oxides for 100 °C annealed samples, as their color remains intact, suggesting the presence of metallic transition metals in general. However, samples annealed at 250 °C showed the absence of C and reduction in oxygen atomic fraction and is ~20 at% only, substantiating the observed undecomposed organic precursors at 100 °C and partial conversion of transition metals into respective oxides. The oxygen atomic fraction enhanced up to 38 at% for 400 °C annealed samples, indicating the increase in transition metal oxide compositions against the unreacted metallic fraction in less temperature annealed samples. In addition to the Co and Mn elements, Fe and Cr are also observed for all samples, explaining the EDX elemental contribution from substrates. The carbon decomposed completely at higher annealing temperatures ~250 and 400 °C. This may be due to the lower decomposition temperatures of initially selected precursors or the content is lower than the detection limits of EDX, used in the present studies. The thicknesses of the coatings were determined from stepped SEM measurements, as illustrated in Fig. 5.

The thickness of these films is around ~1.15 μm and that of SS substrates is ~0.1 mm, as marked in Fig. 5. The optical properties such as absorptivity and emissivity are measured using UV–Vis and Fourier Transform Infrared (FTIR) reflectance measurements in 300–900 nm and 2.5–25 μm wavelength range, respectively. The measured absorptance values for all the samples, annealed at 400 °C are plotted in Fig. 6. We observed that the absorber films prepared from different transition metal precursors, and annealed at 400 °C, as explained in Fig. 6, exhibit α nearly ~0.9, independent of transition metal precursor choice. However, the absorptivity α values increase with annealing temperature, and the measured α values are listed in Table 2.

These results also substantiate that α is nearly independent of the transition metal precursor choice and values are constant for thin film coating structures annealed at the same temperatures. When compared with the α value of bare SS, which is 0.41, those obtained for the coatings are found much higher, the absorption in coating

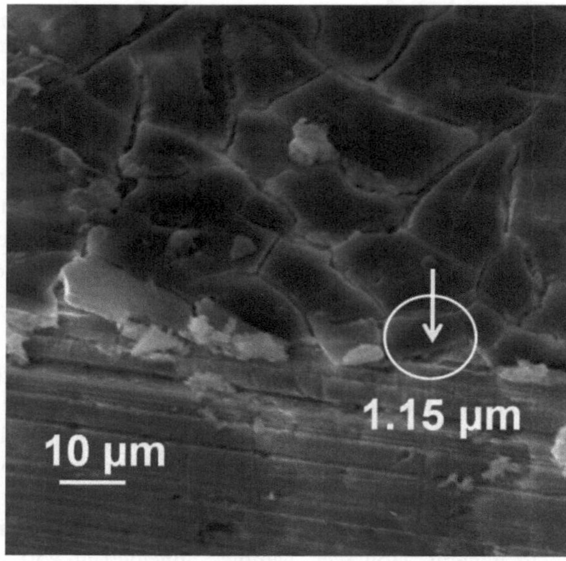

Fig. 5 Scratched surface of a Co and Fe-containing film, heat treated at 250 °C, showing film thickness ~1.15 µm

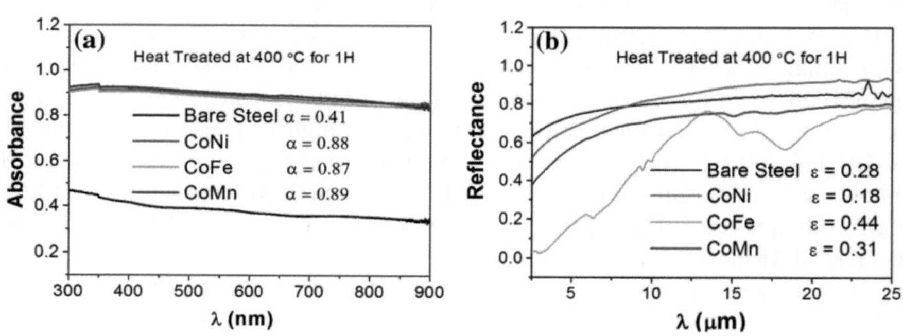

Fig. 6 The absorbance (**a**) and reflectance (**b**) plots for the transition metal absorber layers against wavelengths, and SS substrate

Table 2 Absorptance values for the transition metal oxide coatings when heated at 100, 250, and 400 °C

Coatings	α		
	100 °C	250 °C	400 °C
CoNi	0.64	0.66	0.88
CoFe	0.64	0.74	0.87
CoMn	0.62	0.71	0.89

structures is mainly due to the metal–metal oxides composite structures, formed during controlled oxidation of transition metal precursors. The wavelength range of 300–900 nm is used in these measurements, due to the limitation of the UV–Vis spectrometer operating range.

The observed trend of large values in absorbance within the 300–900 nm regime, may continue till 2500 nm and the integrated absorptance may even be larger than those obtained for these samples. The measured emittance values are plotted in Fig. 5b. For bare SS substrate $\varepsilon \sim 0.28$ and 400 °C annealed CoNi, CoFe, CoMn precursor based coating structures emissivity values are 0.18, 0.44, and 0.31, respectively. The observed ε values are higher than the desired values for spectrally selective coatings, used in solar thermal applications.

This may be due to the absence of the infrared reflecting layer between the substrate and transition metal coating structures. Similar observations of high ε values have also been observed in the case of without infrared reflecting layer for electrodeposited black chrome solar selective absorbers. In black chrome absorber layers, nickel infrared reflector plays an important role to reduce the emissivity ~0.1 or less [11]. Additionally, the values of ε can be tailored further by optimizing the annealing conditions for these structures, causing the controlled oxidation of coated structures, leading to optimized metal–metal oxide based cermet structures, thus providing the desired ε values. The measured UV–Vis normal reflectance values are used to calculate the refractive index, η and extinction coefficient, $\mathfrak{k}$ for the metal–metal oxide thin films in the 300–900 nm wavelength range using the following formulae:

$$n = \frac{100 + \sqrt{R}}{100 - \sqrt{R}}, \tag{1}$$

$$\mathfrak{k} = \frac{A\lambda}{4\pi} \tag{2}$$

where R is the measured reflectance and A is the absorption coefficient of the films. The measured results are summarized in Fig. 7. These measurements suggest that η for bi-transitional metal oxide films are nearly independent at higher wavelength and shows variation near lower wavelengths. The extinction coefficient $\mathfrak{k}$ values for metal oxide films are much larger as compared to bare stainless steel substrate, suggesting the observed high absorptance in these metal–metal oxide thin films structure over the entire 300–900 nm wavelength range. The detailed compositional and elemental analysis will help in understanding the contribution of different components responsible for large extinction coefficient values. The extension of the present work is focused on the development of large-scale coating on specific surfaces and optimization of annealing conditions for desired values of absorptivity and emissivity of the fabricated structures.

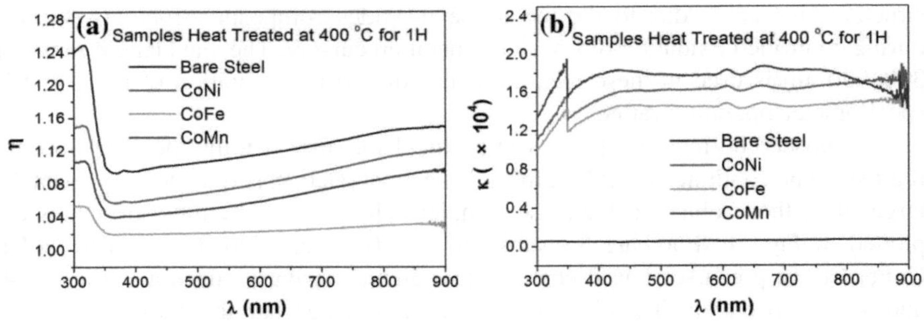

Fig. 7 Plots of **a** refractive index, η and **b** extinction coefficient, k of the films heated at 400 °C against respective wavelength λ in the 300–900 nm range. The abrupt jump at 350 nm in k is because of the grating cross-over during measurements

4 Conclusion

We report the development of mixed transition metal oxide spectrally selective films on SS substrates for solar thermal applications using in-house developed spray coating system. The metal–metal oxide composite absorber layer coatings exhibit high absorptivity ($\alpha \sim 0.9$) in 300–900 nm wavelength range. The thickness of these films is around 1 μm and can be tailored as per requirement. The development of the large-scale system and integration of infrared reflecting layer are in progress to realize the spectrally selective coatings on the desired surface with optimal solar thermal performance such as high absorptivity ($\alpha > 0.9$) and low emissivity ($\varepsilon < 0.1$).

Acknowledgements Authors acknowledge financial assistance from Ministry of New and Renewable Energy (MNRE) through project #15/40/2010-11/ST. Author Ajoy K. Saha acknowledges Rajesh Kumar, Vinod Kumar, Gaurav Kumar, Rakesh Joshi, and Devaiah Soyam for their valuable suggestions and discussions during the work.

References

1. N. Selvakumar, H.C. Barshilia, Sol. Energy Mater. Sol. Cells **98**, 1–23 (2012)
2. C.E. Kennedy, *Review of Mid-to High-Temperature Solar Selective Absorber Materials*. (National Renewable Energy Laboratory, Golden, Colorado, USA 2002), p. 1617
3. L. Kaluža, A. Šurca-Vuk, B. Orel, G. Dražič, P. Pelicon, J. Sol-Gel. Sci. Technol. **20**, 61 (2001)
4. M. Jolya, Y. Antonettia, M. Pythona, M.A. Gonzalez Lazoa, T. Gascoua, A. Hessler-Wyserb, J.-L. Scartezzinia, A. Schülera, Energy Procedia **57**, 487–496 (2014)
5. L. Kaluza, B. Orela, G. Drazic, M. Kohl, Sol. Energy Mater. Sol. Cells **70**, 187–201 (2001)
6. M.-L. Cantúa, A.M. Sabiob, A. Brustengac, P.G. Romeroa, Sol. Energy Mater. Sol. Cells **87** (1–4), 685–694 (2005)

7. M. Voinea, A. Duta, J. Optoelectron. Adv. Mater. **9**(5), 1454–1456 (2007)
8. G.E. McDonald, NASA Technical Memorandum TMX-71596, (1974)
9. M. Madhusudana, H.K. Sehgal, Appl. Energy **10**(1), 65–74 (1982)
10. C.S. Uma, L.K. Malhotra, K.L. Chopra, Bull. Mater. Sci. **8**(3), 385–389 (1986)
11. Y.A. Cengel, Heat and Mass Transfer: A Practical Approach, 3rd edn. (2008)

Part V
Thermal Energy Storage

Thermal Conductivity Enhancement of Myristic Acid Using Exfoliated Graphite for Thermal Energy Storage Applications

Rohitash Kumar, Rakshanda Jakhoria and Ambesh Dixit

Abstract The effect of exfoliated graphite (ExG) with myristic acid phase change material has been investigated to understand the enhancement of thermal conductivity and shape stabilization for the ExG—myristic acid composite system. The thermo-chemically synthesized ExG is added to myristic acid in 5, 10, and 15 wt% and the shape stabilized ExG-myristic acid composite samples are prepared. The heat release, solidification time and latent heat of myristic acid and ExG-myristic acid composites are measured using temperature history (T-history) and Differential Scanning Calorimetric (DSC) measurements, respectively. The measured melting temperature and latent heat of fusion of myristic acid are ~55 °C and ~185 kJ/kg, respectively. Latent heat of ExG-myristic acid composite samples has decreased linearly with increasing ExG weight fraction. T-history measurements suggest the reduction in solidification time for ExG-myristic acid composite samples and ~55% reduction in heat release time is observed for 15% ExG-myristic acid composite sample with respect to pristine myristic acid. The observed reduction in heat release time subtantiates the effective enhancement of thermal conductivity for ExG—myristic acid composite samples.

Keywords Phase change materials · Latent heat · Thermal energy storage · Exfoliated graphite

1 Introduction

Phase change materials (PCMs) offer high energy storage density, isothermal heat storage/release, and good thermal stability against charge/discharge cycles for thermal energy storage applications. Such thermal properties of PCMs make them

R. Kumar (✉) · R. Jakhoria · A. Dixit
Department of Physics and Centre for Energy, Indian Institute of Technology Jodhpur, Jodhpur 342011, Rajasthan, India
e-mail: pg201381005@iitj.ac.in

R. Kumar
Defence Laboratory Jodhpur, Jodhpur 342011, Rajasthan, India

L. Chandra and A. Dixit (eds.), *Concentrated Solar Thermal Energy Technologies*, Springer Proceedings in Energy, https://doi.org/10.1007/978-981-10-4576-9_15

suitable for various applications such as temperature stabilization of temporary shelters, storage/transportation of heat sensitive materials, body comfort in harsh environments, compensating the gap between energy supply and demand, recovery of waste heat, solar heating system, air conditioning system, etc. [1–3]. Among available PCMs, fatty acids ($CH_3(CH_2)_nCOOH$) have been used widely due to their large latent heat, strong chemical stability during heating/cooling cycles, negligible undercooling, noncorrosive behavior, easy availability, economical, and eco-friendliness [4–9]. However, fatty acids have low charge/discharge capability because of their low thermal conductivity. In addition, the leakage, during the phase change from solid to liquid, also restricts their utilization in latent heat thermal energy storage systems (LHTESs) [10]. There are numerous reports on improving the thermal conductivity of fatty acid PCMs. These include the use of metallic fins, carbon fibers, carbon nanotubes (CNTs), nanoparticles, graphene, and carbon foam [10–15]. Recently, porous exfoliated graphite has attracted more attention due to its high porosity with the graphitic connecting network, which provides the enhanced thermal conductivity of composite matrix. The high porosity of thermochemically synthesized ExG provides a large surface to volume ratio, which can accommodate large weight fraction of phase change materials (up to 95%) and thus assist in leakage free shape stabilization of PCMS with enhanced thermal conductivity [16–18]. In this study, we exfoliated bulk graphite using thermochemical process and used ExG to prepare ExG-myristic acid composite samples with the different weight fractions of exfoliated graphite (5, 10, and 15%). The thermal properties of these composite samples are investigated to understand the effect of exfoliated graphite on charge/discharge time, melting temperature and enthalpy of myristic acid.

2 Experiment

Myristic acid (MA) phase change material was procured from Central Drug House (P) Ltd. New Delhi and used without any further purification. The thermal properties of myristic acid were investigated using TA DSC Q10 equipment, such as latent heat of fusion, melting and solidification temperature. Sulfuric acid (99% pure), nitric acid (80% pure), and potassium permanganate were purchased from Fisher Scientific for thermochemical exfoliation of Natural graphite flakes (NGFs). NGFs were procured from Signe Aldrich. The pristine NGFs named as sample A for later discussion. NGFs were immersed in a solution of H_2SO_4, HNO_3, and $KMnO_4$ in the ratio of 1:9:3:0.44 (by weight %) respectively and continuously stirred for ~3 h at ambient temperature to convert the pristine NGFs into intercalated or expandable graphite [19]. The chemically treated product was washed with deionized (DI) water till pH reduced to ~6–7 and the resulted solution was filtered to separate the chemically modified graphite flakes. Further, this chemically modified graphite powder was dried in an oven at 90 °C for 5 h and named as sample B hereafter. This sample was thermally treated at ~950 °C for 60 s for realizing exfoliation and the exfoliated sample is named as sample C hereafter.

The complete exfoliation process is explained in the flow chart, where thermo-chemically expanded graphite is shown in the right panel in Fig. 1.

Liquid myristic acid (MA) was mechanically mixed with thermochemically synthesized ExG using vertical mixing arrangement as shown in Fig. 2. MA was mixed in the liquid phase, maintaining ~75 °C temperature of the base platform during the complete mixing process. The mechanical mixing assisted the uniform impregnation of MA into the porous exfoliated graphite matrix to prepare shape stabilized PCM-ExG composites with high thermal conductivity. The ExG-MA composite matrix was prepared using 5, 10 and 15 wt% of ExG to understand the effect of ExG on thermal conductivity and heat release process of the composite systems.

The large porosity of ExG is important to impregnate the maximum weight fraction of phase change materials, where capillary forces between liquid MA and pores in ExG help to provide the stable leak free compact shaped composite systems [15]. The unmagnified photographs are shown in Fig. 2b–d for pristine MA, ExG, and ExG-MA, with 10% ExG by weight, composite systems, respectively.

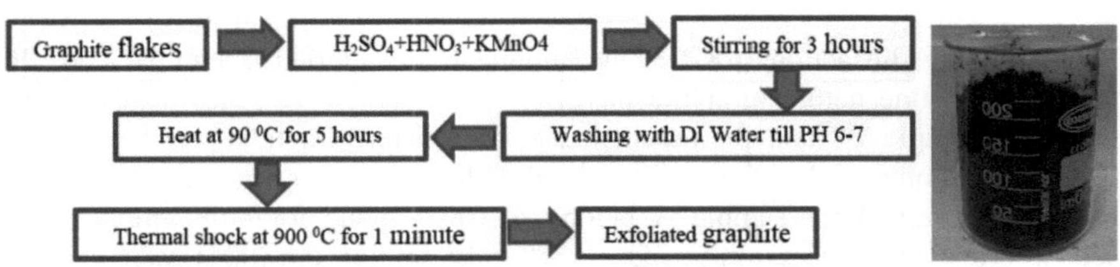

Fig. 1 Process schematic of natural graphite flake exfoliation, with expanded graphite (*right panel*) in beaker

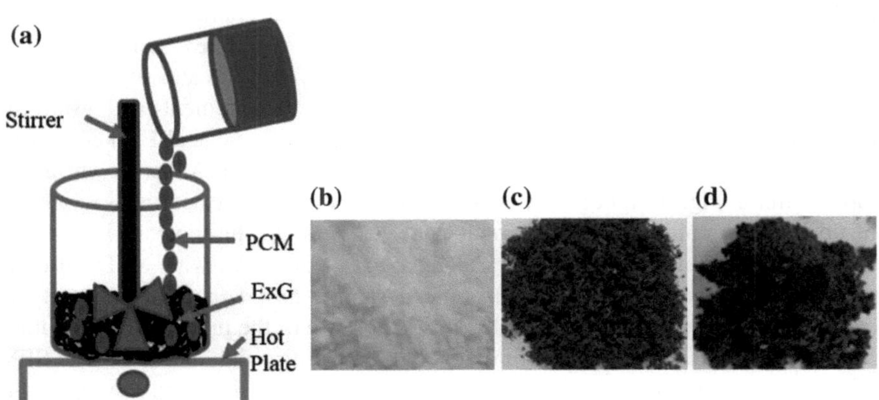

Fig. 2 **a** Schematic of experimental setup for preparation of PCM-ExG composite, **b** photograph of MA, **c** photograph of ExG and **d** photograph of PCM-ExG composite

The ExG-MA composite photograph, Fig. 2d, suggests that MA is held completely inside ExG and trapped inside the pores even in the liquid state. Thus, ExG may also overcome the problem of PCM leakage during thermal charging/discharging of PCM in conjunction with enhanced thermal conductivity. The structural and microstructural properties are investigated using Bruker D8 Advance X-ray diffractometer and Carl Zeiss EVO 18 Scanning Electron Microscope of individual graphite and derivatives, myristic acid, and ExG-MA composite systems. The vibrational frequencies are measured using Bruker vertex 70v Fourier Transform Infrared (FT-IR) Spectroscopy system. The thermal properties, such as latent heat of fusion, melting and the solidification temperatures, are collected using TA DSC Q10 differential scanning calorimetry system. In addition, the heat release experiments are carried out using in-house developed T-history method [22].

3 Results and Discussion

3.1 Structural and Microstructural Properties

Powder X-ray diffraction (pXRD) measurements are carried out in 10°–60° 2Θ range for pristine natural graphite flakes (NGFs) "sample A", chemically treated graphite "sample B" and thermochemically exfoliated graphite "sample C" and results are summarized in Fig. 3. The sharp pXRD reflection peak at $2\Theta \sim 26.4°$, has been observed for sample A belong to (002) plane for bulk graphite with second-order XRD reflection at $2\Theta \sim 55°$, representing the (004) reflection. All these XRD peaks are observed for samples B and C, where the intensity of these diffracted peaks reduced drastically, suggesting that large enhancement in the surface to volume ratio for these samples. This is because of enhanced crystal defects and reduced crystallinity after intercalation and thermal expansion of NGF [19]. In addition, the presence of these XRD peaks substantiate the presence of three-dimensional graphitic structure in expandable graphite "sample B" and exfoliated graphite "sample C" after thermochemical treatment [19]. The microstructural properties are investigated using SEM measurements and results are plotted in Fig. 3 (right panel) for samples A, B, C and myristic acid-ExG composite "sample D" respectively. The compact edges of NGF sample A, as shown in first top column in Fig. 3, right panel, open up after chemical intercalation, as explained in first bottom column in Fig. 3, right panel, suggesting the increase in distance between inter-planer graphite layers along c-axis, consistent with XRD observations, where intensity of chemically treated sample has reduced significantly as compared to the pristine NGF sample. This is because of the intercalation of sulfate and nitrate ions between basal layers of NGFs, in the direction of c-axis. The SEM micrograph of thermally exfoliated graphite, sample C, is shown in the top of the right column of Fig. 3 (right panel), explaining the wormlike structure and these observations are in agreement with the reported literature [19].

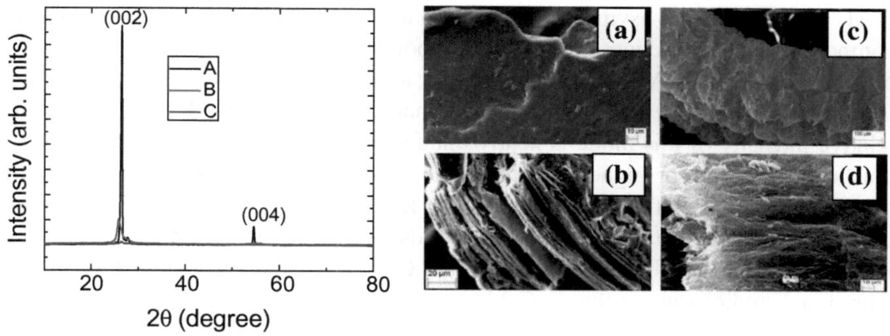

Fig. 3 **a** Powder X-ray diffraction graphs for *samples A, B* and *C*, **b** SEM micrographs of *samples A, B, C* and *D*

Fig. 4 FTIR spectra of MA and ExG-MA composites with 5, 10 and 15 wt% of ExG

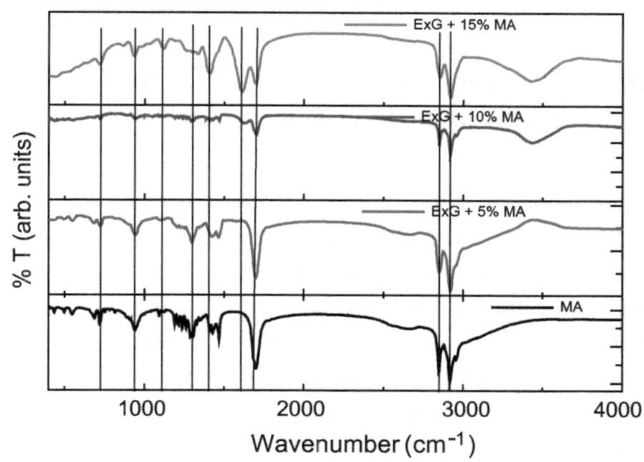

These wormlike structures, Fig. 3b(C), with diamond-shaped pores suggest the enhanced porosity in the final thermochemically treated ExG samples, where edges are connected through pores. The 10% ExG (by weight) ExG-MA composite SEM micrograph, sample D, is shown at the bottom of right side column of Fig. 3, suggesting that MA is impregnated in exfoliated graphite pores. We also carried out FT-IR spectroscopic measurements to understand the chemical interaction between ExG and myristic acid in composite samples. The room temperature FTIR measurements are plotted in Fig. 4 including myristic acid and the different weight percent of ExG-MA composite samples. The chemical formula of myristic acid is $CH_3(CH_2)_{12}COOH$, which consists of C–H symmetric and asymmetric vibrational modes at 2910 and 2850 cm^{-1}, respectively [20]. MA also includes C=O stretching vibrations at 1700 cm^{-1} and out of the plane and in plane swing vibrations of −OH functional group at 935 and 720 cm^{-1}, respectively. All these vibrational modes are observed in the measured FTIR spectrum of myristic acid and different weight percent ExG-MA composite samples, as shown in Fig. 4.

C–H symmetric and asymmetric vibrational modes are unaffected in ExG-MA composite systems, suggesting that C–H bonds are not participating in bonding between ExG and MA. In contrast, the vibration mode at 1700 cm^{-1}, representing C=O stretching mode shows significant changes after adding ExG in MA. The relative intensity of this mode is nearly unaffected in 5% ExG-MA sample, whereas drastically reduces for higher weight fraction of ExG in ExG-MA composite. This may be due to the partial bonding of MA in ExG-MA composite, where C–C bonds in ExG are getting attached with C=O bonds from MA. The vibrational modes at ~1300 and 1410 cm^{-1} are associated with C–C vibrational modes in ExG. These vibrational studies suggest that ExG-MA composites are physically bonded due to the capillary forces between ExG pores and MA and not chemically associated with each other. The strong physical bonding provides mechanical strength to hold the MA phase change material in ExG pores even in melting conditions of MA, avoiding any leakage in case of 10 and 15% ExG by weight in ExG-MA composite systems.

3.2 *Thermal Properties of MA and EXG-MA Composite Systems*

The enthalpy and melting temperature of MA are estimated using Differential Scanning Calorimetry measurements and measured heat flow as a function of temperature are shown in Fig. 5. The 5–10 mg MA sample was weighed and sealed in an aluminum pan using a crimping machine. A sealed empty aluminum pan was used as a reference pan for DSC measurements. Both pans were kept inside DSC cell under continuous nitrogen flow at 50 ml/min to ensure the inert atmosphere inside DSC cell during data collection. The heat flow measurements were carried out at 2 °C/min scan rate from 45 to 60 °C, and measured results are summarized in

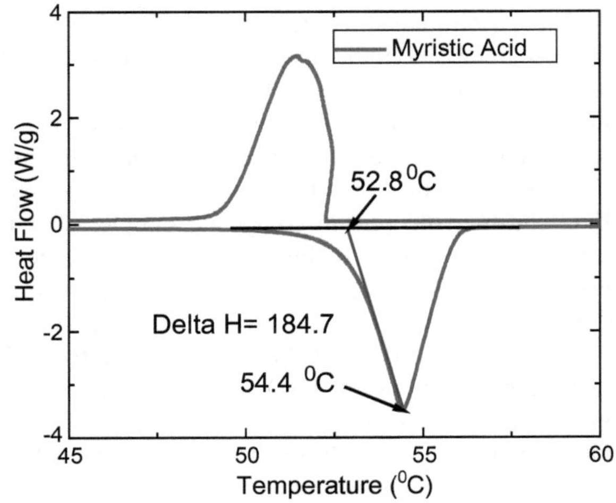

Fig. 5 Differential scanning calorimetric measurement of Pristine MA at heating/cooling scan rate of 2 °C/min

Table 1 Thermal properties of MA

S. No.	Thermal properties	Values
1	Melting temperature	52.8–55 °C[a]
2	Solidification temperature	52.3–50 °C[a]
3	Latent heat	~185 kJ/kg[a]
4	Thermal conductivity	~0.2 W/m-K[b]
5	Density	~0.85 g/cm3[b]

[a]Measured value
[b]Literature value [4]

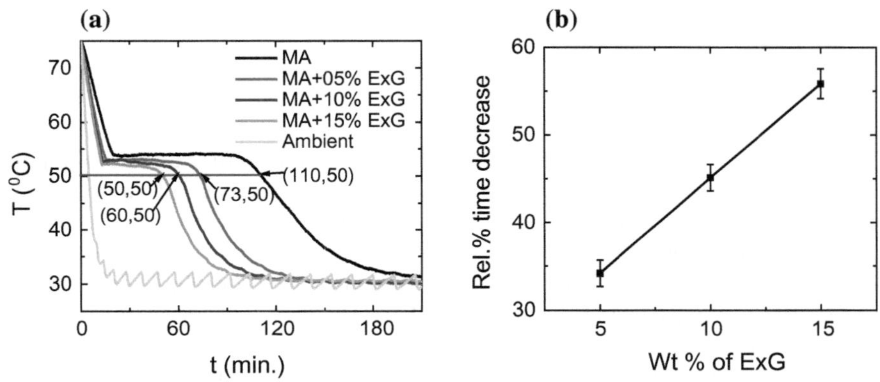

Fig. 6 **a** Cooling response of Pristine MA, MA + 5% ExG, MA + 10% ExG and MA + 15% ExG and **b** reduction in heat release time from 70 to 50 °C for, MA + 5% ExG, MA + 10% ExG and MA + 15% ExG (Color figure online)

Fig. 5. The measurement suggests that melting temperature of MA lies in the range of 52.8–55 °C and solidification temperature lies in the range of 52.3–50 °C for pristine MA phase change material. The slow heating/cooling scan rate was used to ensure the thermodynamic equilibrium during these measurements. The measured latent heat of MA is ~185 kJ/kg and is in agreement with the literature values [21]. These themophysical properties are summarized in Table 1.

Further, to understand the effect of ExG in ExG-Ma composite systems, temperature history (T-History) measurements are carried out for pristine MA and ExG-MA composite samples using in-house developed T-history measuring setup [22]. The weight of the samples was kept constant 20 g and filled in 15 mm diameter and 150 mm length glass tube. The samples were heated up to 75 °C in a heating chamber, ensuring MA is completely charged and is in a liquid state. The cooling response, temperature of samples as a function of time, was recorded at the time interval of every 10 s, in conjunction with chamber temperature, which was set 30 °C during cooling. The measured temperature versus time response for pristine MA and 5, 10, and 15% ExG by weight ExG-MA samples is plotted in Fig. 6a.

We observed that pristine MA and 5, 10, and 15% ExG by weight ExG-MA samples took 110, 73, 60, and 50 min to cool from 70 to 50 °C, respectively.

These measurements explain the reduction in heat release time for ExG-MA composite samples as compared to the pure MA sample. The average reduction in heat release time as a function of ExG weight fraction is plotted in Fig. 6b with the respective error bars, suggesting that 15% ExG by weight-MA composite sample took ~55% less time as compared to the pure MA sample. This is because of ExG network in ExG-MA composite samples, providing a continuous network for thermal transport across the graphitic basal planes. This reduced time is an indirect evidence of thermal conductivity enhancement of ExG-MA composite samples as compared to the pure MA sample.

4 Conclusion

Thermochemical process has been used for the synthesis of exfoliated graphite "ExG" from natural graphite flakes "NGFs" and an increase of large volume enhancement, ~250 times, has been observed. This exfoliated graphite was used to prepare the ExG-MA phase change material composite samples with a different weight fraction of ExG. Thermal response was investigated using in-house developed T-history measurement setup and observed that discharging time of MA reduced significantly by adding ExG. The 15% ExG-MA sample showed ~55% reduction in heat release time as compared to pure MA sample, suggesting an enhancement in thermal conductivity of ExG-MA composite samples. In addition, ExG-MA composite samples exhibit leakage free shape stabilized composite systems for 10% or higher weight fraction of ExG in ExG-MA composite samples. Thus, ExG impregnated phase change material composite systems can be used as thermal energy storage system with enhanced thermal conductivity with reducing leakage problems.

Acknowledgements Authors acknowledge Defence Research and Development Organization (DRDO) through project #CHESS/CARS/ST/004 and Ministry of New and Renewable Energy (MNRE) support through project #15/40/2010-11/ST for financial assistance to carry out the experimental work.

References

1. A. Abhat, Low temperature latent thermal energy storage system: heat storage materials. Sol. Energy **30**, 313–332 (1983)
2. I. Dincer, M.A. Rosen, *Thermal Energy Storage Systems and Applications* (Wiley, Chichester, 2002)
3. B. Zalba et al., Review on thermal energy storage with phase change: materials, heat transfer analysis and applications. Appl. Therm. Eng. **23**, 251–283 (2003)
4. S. Himran et al., Characterization of alkanes and paraffin waxes for application as phase change energy storage medium. Energy Sources **16**, 117–128 (1994)

5. D. Rozanna et al., Fatty acids as phase change materials (PCMs) for thermal energy storage: a review. Int. J. Green Energy **1**, 495–513 (2004)
6. D. Feldman et al., Fatty acids and their mixtures as phase change materials for thermal energy storage. Solar Energy Mater. **18**, 201–216 (1989)
7. A. Hasan, A.M. Sayigh, Some fatty acids as phase change thermal energy storage materials. Renew. Energy **4**, 69–76 (1994)
8. A. Sari, K. Kaygusuz, Some fatty acids used for latent heat storage: thermal stability and corrosion of metals with respect to thermal cycling. Renew. Energy **28**, 939–948 (2003)
9. F. Cedeno, Measurements of temperature and melting heat of some pure fatty acids and their binary and ternary mixtures by differential scanning calorimetry. Thermochim. Acta **369**, 39–50 (2001)
10. B. Zalba et al., Review on thermal energy storage with phase change: materials, heat transfer analysis and applications. Appl. Therm. Eng. **23**(3), 251–283 (2003)
11. S.A. Khateeb, M.M. Farid, J.R. Selman, S. Al-Hallaj, Design and simulation of a lithium-ion battery with a phase change material thermal management system for an electric scooter. J. Power Sources. **128**, 292–307 (2004)
12. L.F. Cabeza et al., Heat transfer enhancement in water when used as PCM in thermal energy storage. Appl. Therm. Eng. **22**, 1141–1151 (2002)
13. J. Fukai et al., Thermal conductivity enhancement of energy storage media using carbon fibers. Energy Convers. Manag. **41**, 1543–1556 (2000)
14. J. Fukai et al., Effect of carbon fiber brushes on conductive heat transfer in phase change materials. Int. J. Heat Mass Transf. **45**, 4781–4792 (2002)
15. H. Mehling et al., Latent heat storage using a PCM–graphite composite material, in *Proceedings of TERRASTOCK 2000*, Stuttgart, Germany (2000)
16. X. Py et al., Paraffin/porous-graphite-matrix composite as a high and constant power thermal storage material. Int. J. Heat Mass Transf. **44**, 2727–2737 (2001)
17. L. Cabeza et al., PCM–graphite matrix in flat plate encapsulates for low temperature applications, in *Proceedings of Future stock 2003, 9th International Conference on Thermal Energy Storage*, Warsaw, Poland (2003)
18. A. Mills et al., Thermal conductivity enhancement of phase change materials using a graphite matrix. Appl. Therm. Eng. **26**, 1652–1661 (2006)
19. Zhang Shengtao et al., Characterization of exfoliated graphite prepared with the method of secondary intervening. Int. J. Ind. Chem. **2**(2), 123–130 (2011)
20. J. Huang et al., Form-stable phase change materials based on eutectic mixture of tetradecanol and fatty acids for building energy storage: preparation and performance analysis. Materials **6**, 4758–4775 (2013)
21. Ahmet Sari, Eutectic mixtures of some fatty acids for low temperature solar heating applications: thermal properties and thermal reliability. Appl. Therm. Eng. **25**, 2100–2107 (2005)
22. R. Kumar et al., An experimental set-up for measuring thermodynamic response of low temperature phase change materials, in *2016 IEEE First International Conference on Control, Measurement and Instrumentation (CMI)*, pp. 107–109 (2016). https://dx.doi.org/10.1109/CMI.2016.7413720

5. [illegible] Paraffin waxes as phase [illegible] material (PCM) for thermal energy storage [illegible] J. Green Energy [illegible]
6. D. [illegible] et al., [illegible] as phase change material [illegible] Sol. Energy Mater. 18, [illegible]
7. [illegible] Some [illegible] thermal [illegible] energy [illegible]
8. [illegible] Some [illegible] latent heat [illegible] corrosion of metals with [illegible] Renew. Energy 28 [illegible]
9. [illegible] Measurements of [illegible] melting heat [illegible] and [illegible] scanning calorimetry. Thermochim. Acta [illegible]
10. G. Zhang [illegible] storage with phase change [illegible] analysis and application. Appl. [illegible]
11. S.A. Khateeb, [illegible] battery with a phase change [illegible] J. Power Sources 128, [illegible]
12. [illegible] Heat transfer [illegible] thermal energy storage. Appl. Therm. Eng. 22, [illegible] (2002)
13. J. Fukai et al., Thermal conductivity enhancement of energy storage media using carbon fibers. Energy Convers. Manag. [illegible] (2000)
14. [illegible] et al., Effect of carbon [illegible] conductive [illegible] phase change materials. Int. J. Heat Mass Transf. [illegible]
15. [illegible] Thermal [illegible]
16. X. Py et al., Paraffin/porous-graphite-matrix composite as a high and constant power thermal storage material. Int. J. Heat Mass Transf. 44, 2727–2737 (2001)
17. [illegible] graphite [illegible] phase change [illegible] application [illegible] Poland [illegible]
18. A. Sari [illegible] Heat [illegible] conductivity [illegible] Appl. Therm. Eng. 26, [illegible] (2006)
19. Zhang [illegible]
20. [illegible] et al. [illegible] building energy [illegible] (2014)
21. [illegible] applications [illegible] properties [illegible] Appl. Therm. Eng. [illegible] (2007)
22. [illegible] et al. [illegible] experimental [illegible] measuring the [illegible] of low [illegible]

Performance Investigation of Lab-Scale Sensible Heat Storage System

Chilaka Ravi Chandra Rao, Hakeem Niyas, Likhendra Prasad and Muthukumar Palanisamy

Abstract This paper presents the theoretical investigation of heat storage characteristics and transient behaviour of a sensible heat storage (SHS) module of 10 MJ storage capacity designed for discharging the heat in the temperature range of 523–623 K for solar power plant applications. Thermal model of heat storage module in cylindrical configuration has been developed considering the heat transfer enhancement technique in the storage module by incorporating the axial fins on the discharging tube surfaces. High thermal conductivity (cast iron and cast steel) and low thermal conductivity (concrete) materials have been chosen as the SHS materials for the present analysis. Number of discharging tubes with axial fins over the tube periphery has been optimized based on the charging time.

Keywords Sensible heat storage · Thermal modelling · Concrete · Cast iron · Cast steel

Nomenclature

a	Centre distance between adjacent tubes, (m)
b	Thickness of fins on the HTF tubes, (m)
C_{ps}	Specific heat of SHS material, (J/kg K)
C_{pf}	Specific heat of heat transfer fluid, (J/kg K)
d	Internal diameter of the HTF tubes, (m)
D	Diameter of storage module, (m)
h	Height of fins on the charging tubes, (m)
k_s	Thermal conductivity of SHS material, (W/m K)
L	Length of SHS module, (m)
m	Mass of SHS material, (kg)
n	Number of HTF tubes
n_{fin}	Number of fins on a HTF tube

C.R.C. Rao · H. Niyas · L. Prasad · M. Palanisamy (✉)
Department of Mechanical Engineering, Indian Institute of Technology Guwahati, Guwahati, Assam, India
e-mail: pmkumar@iitg.ernet.in; pmkumariitg@gmail.com

L. Chandra and A. Dixit (eds.), *Concentrated Solar Thermal Energy Technologies*, Springer Proceedings in Energy, https://doi.org/10.1007/978-981-10-4576-9_16

Q	Heat storage capacity, (J)
t	Discharging time, (s)
t_{eff}	Effective discharging time, (s)
T_{ini}	Initial temperature of storage system, (K)
T_{inlet}	HTF inlet temperature, (K)
T_{outlet}	HTF outlet Temperature, (K)
V	Volume of storage material, (m^3)
V_{min}	Minimum volume of storage material required to store 10 MJ, (m^3)
ρ_s	Density of solid-state SHS material, (kg/m^3)
ρ_f	Density of HTF, (kg/m^3)
μ	Dynamic viscosity of HTF, (Ns/m^2)
η_{disch}	Discharging energy efficiency
$\vec{v}$	Velocity of HTF, (m/s)

1 Introduction

Energy saving and development of efficient energy storage systems have been the main objectives especially when the source of energy is intermittent like solar energy. Integration of thermal energy storage (TES) system can enable a constant power generation from solar thermal power plants (STPP). The three main types of TES systems are sensible heat storage (SHS), latent heat storage (LHS) and thermo-chemical heat storage (TCHS) systems. In SHS systems, heat is stored by raising the temperature without causing the phase change of storage medium (liquid or solid). Liquid media (mostly molten salts) is the proven technology for SHS systems but major problems with liquid media system are bulky storage tanks for hot and cold fluids and expensive heat exchangers. Although one tank can be eliminated by using thermocline system [1] but freezing of molten salts at high temperature is still a key problem and the requirement of auxiliary heating units during freezing period leads to higher operating cost. In addition, most of the molten salts become unstable beyond 550 °C [2]. An alternative to liquid media storage is LHS that involves phase transition (i.e. solid to liquid and vice versa) of storage material for storing/releasing heat. LHS technology is in development stage and presently, no commercially large-scale storage heat applications are available.

Fernandez et al. [3] and Khare et al. [2] have shown that concrete, cast iron, and cast steel can be the suitable materials of high temperature SHS application. The TES system using solid-state SHS material is generally employed by set in a multiple tube heat exchanger in the SHS material to transfer the heat from/to the heat transfer fluid (HTF), such as air, molten salt, steam, and thermic oil. The benefits of concrete-based storage systems are the low storage material costs and low degradation of heat transfer between the HTF and concrete [4]. The main requisite properties of SHS material are high heat capacity, density and heat conductivity. Tamme et al. [5] explored the possibility of employing ceramic and concrete as SHS materials for high temperature heat storage applications. Laing et al. [6] performed the investigation on the SHS system employing ceramic and concrete for 350 kW storage capacity at 663 K. Concrete is selected due to its low exergy loss, low cost and easy handling although ceramic has 35% more thermal conductivity and 20% higher storage capacity. Laing et al. [7] presented the design and test results of a SHS system employing concrete as SHS material and thermic oil as HTF in the temperature range of 573–673 K. For a storage capacity of 400 kWh, the reported charging and discharging times were 6 h. According to John et al. [8], concrete has better resistance to thermal cyclic loading during charging and discharging and could retain its mechanical properties over such cycles. The major issues associated with concrete as a SHS material are the larger size and slower heat transfer rate. However, heat transfer characteristics in low conductivity solid storage media such as concrete, magnesia, etc., can be improved by incorporating heat transfer enhancement techniques [9] and storage module size can be minimized by selecting high thermal capacity materials with high temperature swing. Nandi et al. [10] reported that concrete and castable ceramic are less cost (25–30$ kWh) and durable SHS materials. Sragovich [11] analysed the transient behaviour of SHS system using magnesia as SHS material. It was observed that there always exists a critical mass flow rate of HTF at which the outlet temperature of HTF drops suddenly and this critical value varies in the same fashion as the tube diameter. Miro et al. [12] have tested a by-product of potash industry as a SHS material for its suitability. Anderson et al. [13] predicted the performances of heat storage system employing alumina as the SHS material using compressed air as the HTF.

From the previous reported works, it is seen that most of the researchers [3–5, 7, 8, 10] emphasized the usage of concrete as a SHS material at high temperatures. However, very few works are reported on the heat transfer enhancement studies in solid media storage, although feasibility study has been done. Further, the optimization of the number of HTF tubes used in the solid media storage system is not reported in the literature. It is also important to note that for a given geometry of SHS unit, its heat transfer augmentation techniques also need to be optimized. In the recent studies by the author's research group [14], the charging characteristics of the SHS system employing cast iron, cast steel storage and concrete as the SHS materials were presented. In this study, discharge characteristics of solid-state heat storage systems are analysed employing two high thermal conductivity solid-state SHS materials (cast iron and cast steel) and a low thermal conductivity solid-state SHS material (concrete). Number of HTF tubes and axial fins on the HTF tube is optimized based on the storage module's charging time using COMSOL™ Multi-physics 4.2. The thermal storage performances of the selected SHS materials are predicted for discharging 10 MJ stored heat.

2 Design of Sensible Heat Storage Model

The main aspects considered in the design of storage system are the thermophysical properties and storage temperature range of SHS materials. The design procedures for SHS model are given below:

- Fix the charging temperature range (ΔT_{ch} = 100 K) of storage module.
- Cast iron, cast steel and concrete are selected as SHS materials based on the thermophysical properties, cheaper cost [2, 10] and more availability. Also, the selected SHS materials are modular, i.e. they can be easily implemented for higher heat storage modules. The thermophysical properties of SHS materials, copper and thermic oil are given in Table 1.
- The minimum volume (V_{min}) of SHS material needed for storing 10 MJ of heat is found using Eq. (1).

$$Q = \rho_s V_{\min} C_{ps} \Delta T_{ch} \text{ (J)} \quad (1)$$

- Equation (1) can be rewritten as Eq. (2). The values of D^2L for concrete, cast iron and cast steel are 0.06808, 0.03157 and 0.0272, respectively. Now these values are arrived by trial and error technique for different *L*/*D* ratios as presented in Table 2.

Table 1 Thermo-physical properties of SHS material [4, 16], copper [17] and thermic oil

Sl. No.	SHS/tube/HTF material	Density (kg/m^3)	Thermal conductivity (W/m K)	Specific heat (J/kg K)	Dynamic viscosity (Ns/m^2)
1	Concrete	2200	1.5	850	–
2	Cast steel	7800	40	600	–
3	Cast iron	7200	37	560	–
4	Copper	8933	401	385	–
5	Thermic oil	761	0.121	2800	0.02

Table 2 Selection of diameter (*D*) and length (*L*) of SHS modules

D (m)	D^2 (m^2)	Length (m)				$D^2 * L$(m^3)			
		L/*D*				*L*/*D*			
		1.5	2	2.5	3	1.5	2	2.5	3
0.20	0.04	0.3	0.4	0.5	0.6	0.012	0.016	0.02	0.024
0.25	0.0625	0.375	0.5	0.625	0.75	0.02343	0.03125	0.03906	0.04687
0.30	0.09	0.45	0.6	0.75	0.9	0.0405	0.054	0.0675	0.081
0.35	0.1225	0.525	0.7	0.875	1.05	0.06431	0.08575	0.10718	0.12862
0.40	0.16	0.6	0.8	1	1.2	0.096	0.128	0.16	0.192

$$D^2L = \frac{4Q}{\pi \rho_s C_{ps} \Delta T} \ (\mathrm{m}^3) \tag{2}$$

- It can be observed from Table 2 that the value of D^2L which is close to the required value (0.068087, which is obtained from Eq. 1) of D^2L for concrete is 0.0675 for $D = 0.3$ m and $L/D = 2.5$. Therefore, the chosen concrete module's diameter (D) and length (L) are 0.30 and 0.80 m, respectively. Similar methodology is adopted for cast iron and cast steel modules and the required D^2L will be satisfied with $D = 0.25$ m and $L = 0.6$ m for both the materials.
- Diameter (d) of the tube is selected as 0.0125 m with a wall thickness of 1.5 mm. Half-inch tube diameter is chosen as it is easily available and also the volume available for heat storage with this size of tube is within limit. Selecting the next standard size, i.e. ¾ or 1 inch would not satisfy the minimum volume criteria. SHS material volume at different standard sizes of tube diameters is tabulated in Table 3.
- Volume of SHS material available for heat storage is estimated using Eq. (3) and it is ensured that this volume is not less than the minimum volume obtained from Eq. (1) for all three materials.

$$V = \left[\frac{\pi}{4}\left(D^2 - nd^2\right) - nn_{\mathrm{fin}}bh\right] L \ (\mathrm{m}^3) \tag{3}$$

- Straight rectangular fins with thickness (b) of 2 mm, height (h) 10 mm and length equal to the storage module are chosen.

Table 3 Selection of tube diameter (d) as per standard size availability for SHS concrete module

Diameter of the module, D (m)			0.2	0.250	0.30	0.350	0.4
Length of the module, L (m)	L/D = 1.5		0.3	0.375	0.45	0.525	0.6
	L/D = 2		0.4	0.500	0.60	0.700	0.8
	L/D = 2.5		0.5	0.625	0.75	0.875	1.0
	L/D = 3		0.6	0.750	0.90	1.050	1.2
Available volume of the module, $V \times 10^{-3}$ (m^3)	L/D 1.5	d (1/2″)	8.269	16.96	30.08	48.49	73.09
		d (3/4″)	7.387	15.86	28.75	46.94	71.32
		d (1″)	6.171	14.34	26.93	44.82	68.89
	L/D 2	d (1/2″)	11.03	22.62	40.10	64.65	97.45
		d (3/4″)	9.849	21.15	38.34	62.59	95.10
		d (1″)	8.228	19.12	35.90	59.76	91.85
	L/D 2.5	d (1/2″)	13.78	28.27	50.13	80.82	121.8
		d (3/4″)	12.31	26.43	47.92	78.24	118.9
		d (1″)	10.29	23.90	44.88	74.69	114.8
	L/D 3	d (1/2″)	16.54	33.93	60.15	96.98	146.2
		d (3/4″)	14.77	31.72	57.50	93.89	142.6
		d (1″)	12.34	28.68	53.86	89.63	137.8

Table 4 Estimated mass of SHS material

SHS material	D (m)	L (m)	V_{min} from Eq. (1) (m^3)	V from Eq. (2) (m^3)	m (kg)	m with 20% margin (kg)
Concrete	0.30	0.8	0.0535	0.0535	118	142
Cast iron	0.25	0.6	0.0214	0.029	230	276
Cast steel	0.25	0.6	0.0248	0.029	212	254

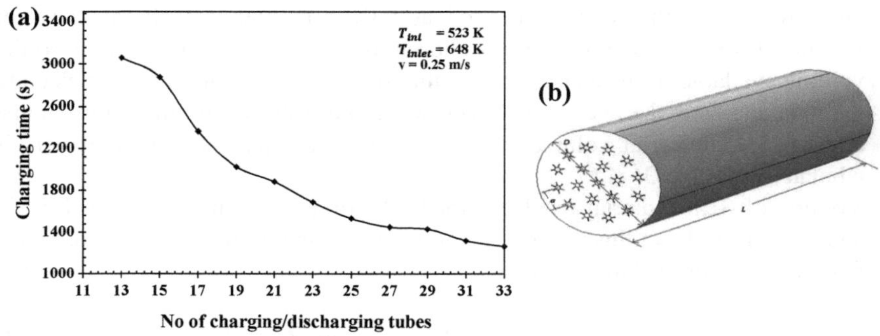

Fig. 1 **a** Optimization of the number of HTF tubes, **b** optimized thermal model of 10 MJ capacity [14]

Storage volume of the module is calculated considering copper fins on the discharging tubes. Based on the above calculation, the mass of SHS material with a factor of safety of 1.2 on actual available mass (obtained from Eq. 3) is given in Table 4.

2.1 Optimization of the Number of HTF Tubes of the SHS Module

Decreasing the number of HTF tubes (n) increases the charging/discharging time considerably but also improves the SHS capacity slightly. Hence, to optimize the number of HTF tubes, the charging times with respect to the different arrangements of HTF tubes are presented in Fig. 1a. From Fig. 1a, it can be seen that there are two ranges of tube arrays at which the charging time does not vary significantly [14]. First range of charging tube array lies between 19 and 21 tubes with the charging time difference of 72 s and second range of 27–29 tubes with the charging time difference of about 20 s. If the second range of tubes is selected, then the charging time of the storage module reduces by 10 min approximately. At the same time, it incurs additional cost of 8–10 tubes and also reduces the effective storage volume below the minimum storage volume (from Eq. 1) required for storing 10 MJ of heat. Thus, as shown in Fig. 1b, the number of HTF tubes in the optimum module configuration is selected as 19.

3 Thermal Modelling of Sensible Heat Storage Module

3.1 Model Description

The storage module acts like a heat exchanger that releases/absorbs heat energy when the cold/hot HTF is passed through the HTF tubes/holes. It consists of a cylindrical module with finned tubes embedded into concrete module (Fig. 1b). For cast iron and cast steel, simply holes/drills are made. Three-dimensional conjugate heat transfer model of cylindrical storage module which includes 3D solid and 3D fluid domains has been considered in the present study. In order to minimize the computational time and cost, one-fourth of the storage module has been selected as the heat transfer is symmetrical (shown in Fig. 2 for concrete module). During discharging of heat from module, HTF is supplied at lower temperature to the storage module from one end and it leaves from the other end. While passing through the tubes, HTF absorbs the heat energy from the charged storage module through the mode of conjugate heat transfer. The details of physical model of 10 MJ SHS system are already reported by the authors [14].

3.2 Governing Equations

In this section, a summary of thermal model employed for simulating the transient behaviour of the SHS module is presented. The assumptions employed in the thermal model are: (i) HTF inlet velocity profile is fully developed, (ii) SHS module material is isotropic and no axial conduction in the HTF, (iii) outer surface of the cylinder is well insulated, and thus it is adiabatic and (iv) neglected radiation effects. Cast iron and cast steel modules have zero contact resistance as these metallic modules have drilled holes and direct heat transfer takes place between

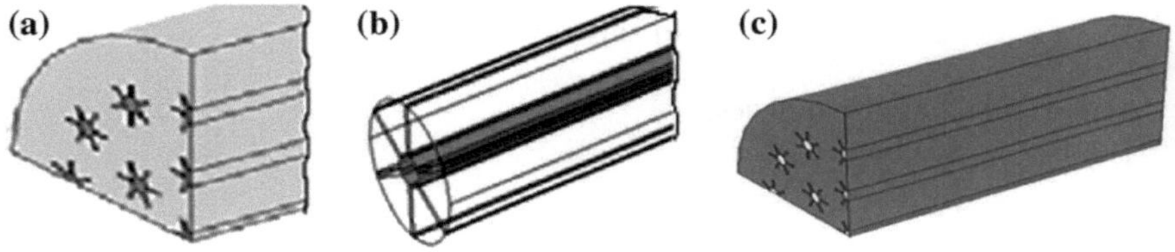

Fig. 2 Initial condition **a** at inlet, **b** no slip boundary condition at interface of a tube and **c** thermal insulation boundary condition [14]

HTF and cast iron or cast steel module. In case of concrete module, the contact resistance is neglected as the charging tubes are buried inside the concrete module.

The governing equations for the fluid flow and heat transfer between the HTF and SHS module are as follows:

Fluid flow

$$\nabla \cdot \vec{v} = 0 \tag{4}$$

$$\rho_f \frac{D\vec{v}}{Dt} = -\nabla P + \mu \nabla^2 \vec{v} \tag{5}$$

Convection: Solid-liquid interface

$$\rho_f C_{pf} \frac{DT}{Dt} = k_s \nabla^2 T \tag{6}$$

Conduction: Solid region

$$\rho_s C_{ps} \frac{\partial T}{\partial t} = k_s \nabla^2 T \tag{7}$$

3.3 *Initial and Boundary Conditions*

Initially, at inlet (t = 0), there is no HTF flow through the tubes (i.e. at rest). All the domains are also initially assigned a constant temperature of T_{ini}. The discharge process is commenced by specifying a constant fluid inlet temperature of T_{inlet} and a constant fluid velocity. Except the inlet and outlet, all the other surfaces are insulated. Figure 2a–c illustrates the initial and boundary conditions of 3D cylindrical storage module model [14, 15].

Inlet:

$$\begin{aligned} &\text{At}\,(t = 0) : \vec{v} = 0; \quad T = T_{\text{ini}} \\ &\text{At}\,(t > 0) : \vec{v} \neq 0; \quad T = T_{\text{inlet}} \end{aligned}$$

Interface:

$$\text{No slip}\,(\vec{v} = 0)$$

At the solid–liquid Interfaces.

Thermal insulation:

$$n \cdot (k \nabla T) = 0;$$

where, n is the normal vector.

3.4 Performance Parameters

The performance parameters are discharging time, effective discharging time (for concrete), energy recovered and discharging energy efficiency. These parameters are useful to evaluate the transient performance of the storage modules during the discharge process. Complete discharging time is the time taken by the storage module's average volume temperature to attain the HTF inlet temperature, T_{inlet}. Effective discharging time is the time at which the temperature difference between the storage module's volume average temperature and the outlet temperature of the HTF becomes 5 °C, i.e. for maintaining better heat transfer rate between the storage module and heat transfer fluid, at least 5 °C temperature difference is recommended due to less thermal conductivity of concrete. The amount of heat retrieved from the storage module is calculated using Eq. (8).

$$Q = \rho_s V C_{ps}(T\text{ini} - T(t)), \tag{8}$$

where T_{ch} is the volume average temperatures of module at the end of charging cycles.

Discharging energy efficiency is the ratio of the energy released during the discharging cycle to the maximum energy that can be recovered from the storage module as applied in case of charging energy efficiency [16]

$$\eta_{\text{disch}} = \frac{T_{\text{ini}} - T(t)}{T_{\text{ini}} - T_{\text{inlet}}} \tag{9}$$

3.5 Grid Independence Test

Comparisons are made for results obtained from different mesh densities for the concrete module configuration to make sure that the results are mesh independent. The range of the densities is varied from 43,969 to 139,438 elements with three different mesh profiles at the predefined temperatures (shown in Fig. 3). The three grid profiles are extremely coarse (43,969 elements), extra coarse (67,667 elements) and coarser (139,438 elements). As illustrated in Fig. 3, the results obtained from the mesh sizes of 67,665 (extra coarse) and 139,438 (coarser) matched closely. It is also noted that increasing the mesh elements beyond coarser mesh (fine mesh 269,000) leads to an increase in computational time by about 30 h. In order to save the computational time, mesh size of 67,665 has been chosen for the further analysis. The governing equations are solved using the time-dependent PARADISO solver. The simulations are carried out using Intel (R) Core™ i5-2400CPU@3.10 GHz machine (installed memory of 8 GB) with a time stepping of 0.1 s.

3.6 Validation of Current Model

To validate the developed thermal model, the obtained numerical results (charging time) of the storage module at different thermal conductivity values are compared with the data reported by Tamme et al. [5]. The initial and boundary conditions, thermophysical property of the SHS module selected for the numerical validation are taken from Tamme et al. [5]. It is understood from Fig. 4 that the numerical results predicted for the current model showed a good agreement with the data reported by Tamme et al. [5]. However, at the initial period of charging, a small deviation was observed which might be due to neglecting the axial conduction in HTF.

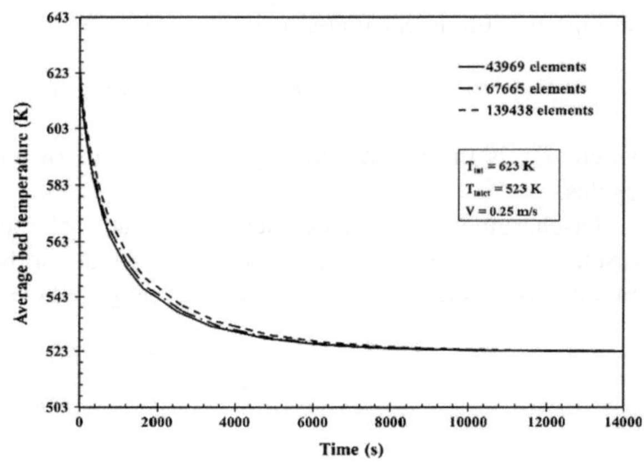

Fig. 3 Grid independent test for the concrete module

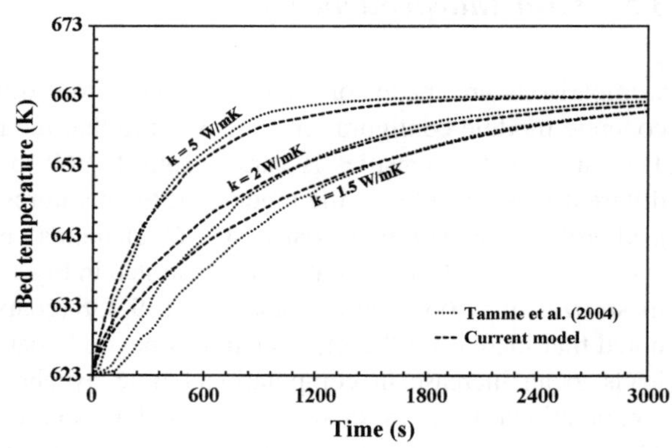

Fig. 4 Validation of the numerical model [14]

4 Results and Discussion

In the following section, some of important numerical results obtained from the simulation are presented.

4.1 Discharging Time

Storage module is said to be discharged completely when its volume average temperature reaches almost the inlet temperature of HTF. Figure 5a shows the complete discharging time of the concrete storage module. The complete discharging time of the charged concrete unit is about 13,965 s. It can be found that the temperature drop of the concrete module is fast for the first 3600 s due to higher potential of conduction. This driving potential decreases gradually due to lower difference in temperature between the HTF and the storage module. As the time progresses beyond 5050 s, the slope of discharging curve becomes flat. Therefore, considering the minimum temperature difference of 5 °C between average storage module's temperature and HTF outlet temperature (required for effective heat

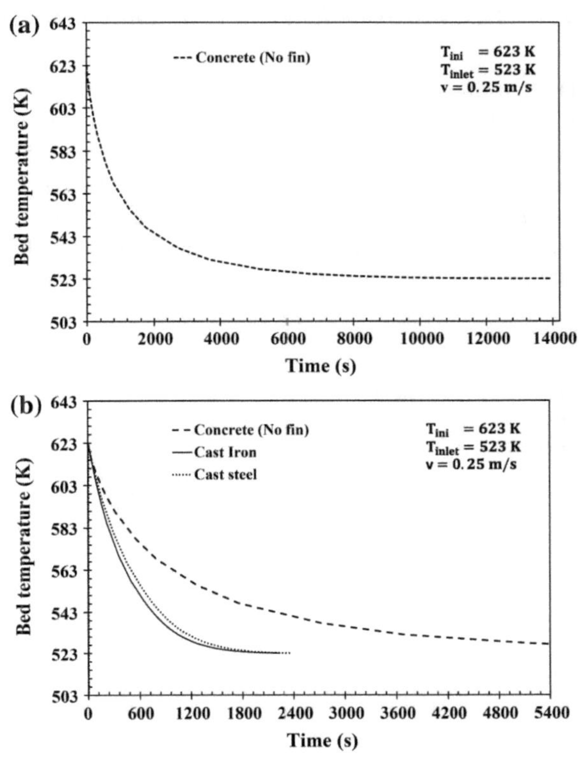

Fig. 5 Discharging time of **a** concrete (complete) and **b** concrete (effective), cast iron and cast steel modules

transfer), 5050 s is considered as the effective discharging time for concrete storage module provided the energy that is recovered is at least 10 MJ. It has been found that the energy recovered at 5050 s is within the limit of the design capacity, which is discussed in next section. The effective discharging time of concrete module and complete discharging time of cast iron and cast steel modules are shown in Fig. 5b. The complete discharging time of the cast iron and cast steel modules is 1728 and 1887 s, respectively.

4.2 Thermal Energy Discharged from the Storage Modules

Figure 6 shows the discharge rate of the thermal energy from the selected SHS materials. The amount of heat discharged from the different SHS modules at their respective discharging times (i.e. effective time for concrete module and complete time for metallic modules) is calculated using Eq. (8). Thermal energy discharged from concrete, cast iron and cast steel storage modules are 12.11, 14.24 and 16.51 MJ, respectively within 5050, 1728, and 1887 s.

4.3 Discharging Energy Efficiency (DEE)

Discharging energy efficiencies of the three selected storage materials are shown in Fig. 7. For the cast iron and the cast steel storage materials, DEE is almost about 100% because the average volume temperature of these storage modules attains the value of the T_{inlet} within their respective discharging times. While the concrete storage module is being less heat conductive, its volumetric temperature does not reach the T_{inlet} within the discharging time of 5050 s. The discharging energy efficiency of the concrete storage module is about 95%.

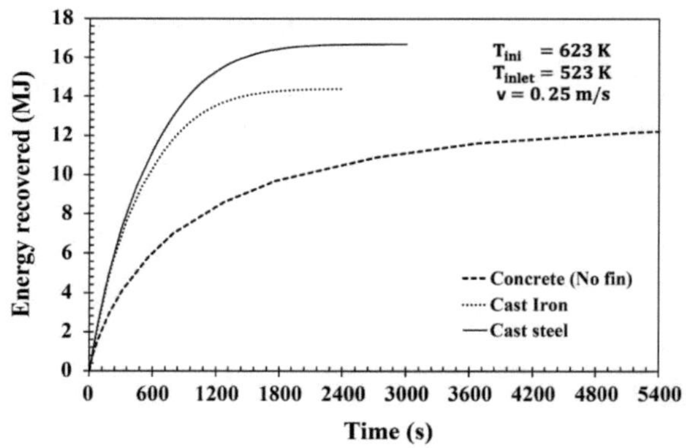

Fig. 6 Thermal energy discharge rate from SHS modules

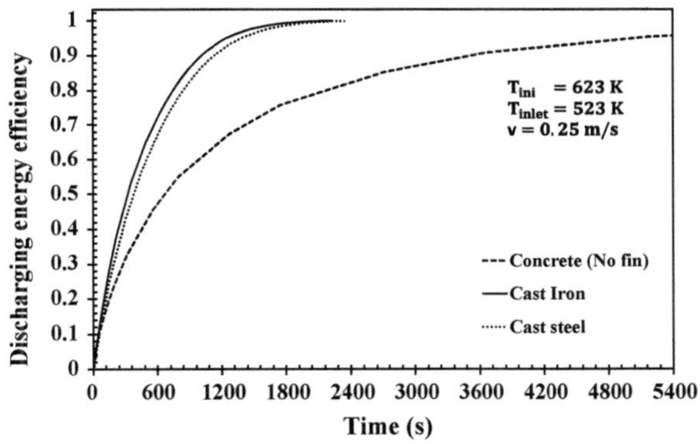

Fig. 7 Discharging energy efficiency of SHS modules

4.4 Axial Variation of HTF Temperature

Initially, the storage module is fully charged and its average volume temperature is 623 K. When the discharging process is initiated by supplying HTF at temperature 523 K, the heat stored in the storage module is transferred to the discharging tube, which is further carried away by HTF through convection. As the HTF moves along the module (axial direction), its temperature rises and exits at a temperature much higher than its inlet temperature. In order to analyse the axial variation of HTF temperature, temperature at an interval of every 10 cm along the axial direction is predicted. It can be noticed that the HTF temperature rise is rapid for the early period of discharging process and after a certain time no appreciable temperature rise is seen as the storage module temperature itself decreases with time resulting into the retardation of driving potential for conduction. The rise in HTF temperature is more significant in cast iron and cast steel as compared to concrete module. For example, the rise in HTF temperature predicted in concrete, cast iron and cast steel module, during the discharging time of 1 min are about 21, 63 and 66 K, respectively. These effects are shown in Fig. 8a–c, for cast iron, cast steel and concrete, respectively. For all the three storage materials, the HTF temperature rise is significant up to 30 min.

4.5 Effect of Fins on Discharging Time

For achieving better heat transfer rate, the effective thermal conductivity of the concrete is enhanced by adding the fins on the outer surface of charging/discharging tubes. In order to account the effect of fins, the discharging time of concrete storage module without fins is compared with finned tube concrete storage module. It is noticed from Fig. 9 that the incorporation of fins on tubes causes the considerable enhancement in heat conduction in the concrete storage module and hence the total discharging time of the storage module is reduced significantly. The reduction in

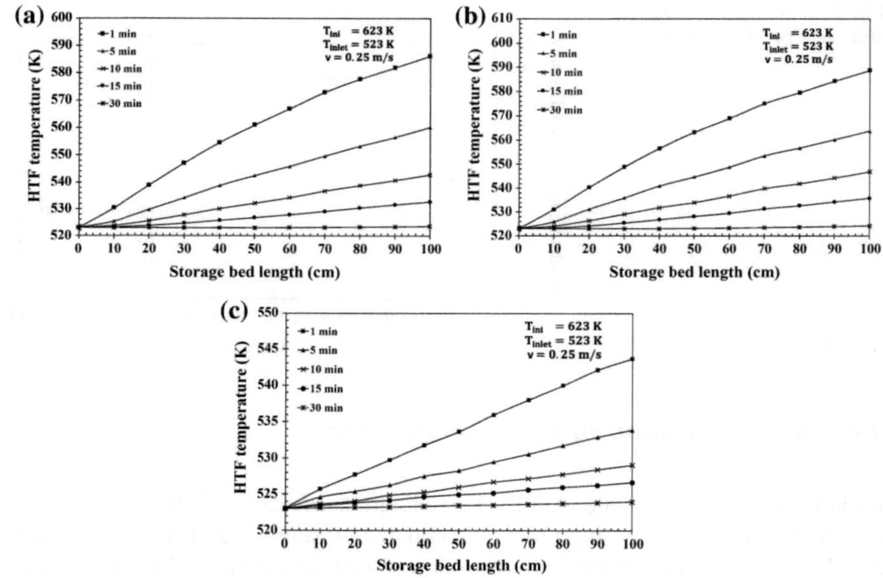

Fig. 8 Axial variation of HTF temperature in the storage module at different time intervals **a** cast iron, **b** cast steel and **c** concrete (no fin)

Fig. 9 Effect of number of fins on discharging time of concrete module

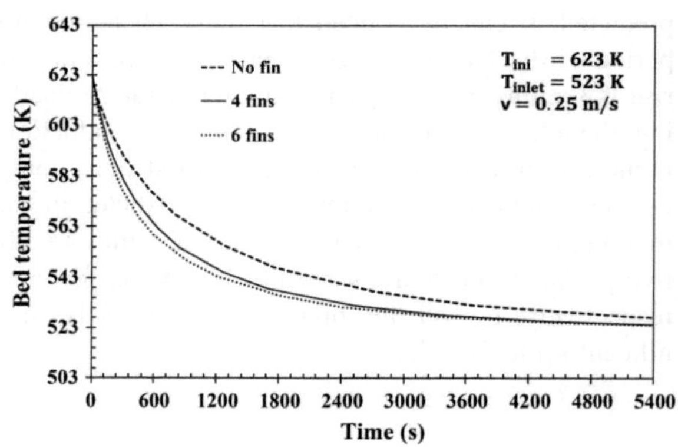

effective discharging times are about 32.2 and 35.9% for the tubes with four fins and six fins cases in comparison with the concrete module without fin. Further increasing the number of fins would not satisfy the minimum volume criteria for required heat storage and it is also uneconomic. Thus for discharging cycle, tube with four fins gives the optimum discharging time.

4.6 Effects of Fins on Energy Recovered

Figure 10 shows the effect of the number of fins on the thermal energy recovered from the storage modules. Although the addition of fins on tubes reduces the volume of concrete marginally but the reduction in the discharging time is very significant. With addition of fins on a discharging tube of concrete module, the reductions in energy discharged from concrete module are 1.7 and 2.6% for four fins and six fins, respectively. Table 5 provides the effect of number of fins on the complete and effective discharge times of the concrete module and the respective heat discharge capacities.

4.7 Effect of HTF Velocity on Discharging Time

Increasing the HTF velocity causes the improvement in the overall heat transfer coefficient, which accelerates the rate of temperature drop of storage module. Hence, it takes less time to reach the T_{inlet} resulting in the reduction of the discharging time. The variation in HTF inlet velocity on the discharging time for concrete with four fins, cast iron and cast steel are shown in Fig. 11a, b. The

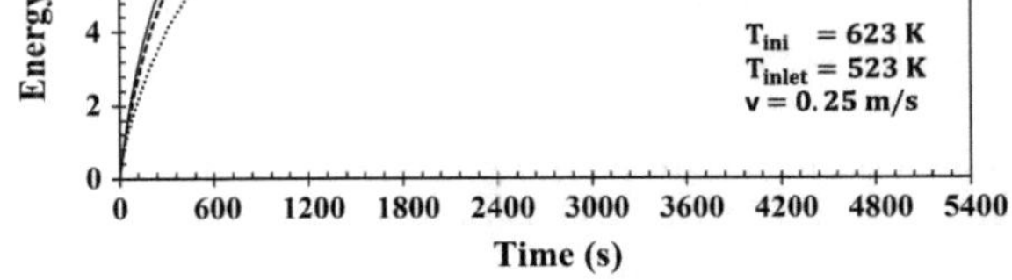

Fig. 10 Effect of number of fins on energy recovered from concrete module

Table 5 Effect of the number of fins on the complete and effective discharge times of the concrete module and the respective heat discharge capacities

n_{fin}	Discharging time (s)		Energy discharged (MJ)		Corresponding storage module temperature (K)
	t_{eff}	t	Effective energy discharged	Complete energy discharged	
0	5050	13,965	12.11	12.74	528
4	3423	9649	11.87	12.48	528
6	3238	9165	11.71	12.31	528

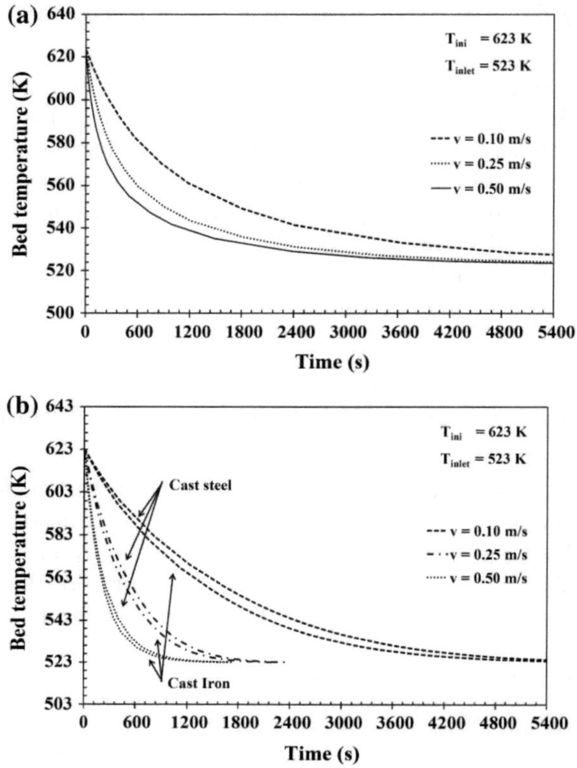

Fig. 11 Effect of HTF velocity on the discharging time of the SHS modules **a** concrete, **b** cast steel and cast iron

effective discharging time of concrete (with four fins) SHS module for velocity of 0.1, 0.25 and 0.5 m/s are 5334, 3423 and 2435 s, respectively. Similarly, the complete discharging time of cast iron and cast steel modules for the corresponding velocities are 5145, 1728 and 1110 s, and 5626, 1887 and 1197 s respectively. The reduction in the discharging time of the SHS module with the case having a HTF velocity of 0.25 m/s when compared to the case with 0.1 m/s is significant but same is not true for the case with a velocity of 0.25 m/s when compared to the case with 0.5 m/s. Further increase in the HTF fluid velocity will not help in reducing the discharging time. Therefore, HTF velocity considered for the current study is 0.25 m/s.

5 Conclusions

Simulated results of SHS system during discharging cycle are presented for three storage materials, viz. concrete, cast iron and cast steel. Heat transfer enhancement technique is implemented by adding fins on the HTF tubes. The effect of improvement in thermal conductivity of storage module with addition of axial fins on the surface of HTF tubes is analysed for three cases, i.e. tubes with no fins, tubes with four fins and six fins. It is found that the concrete with four fins provided the optimum discharging time with acceptable storage capacity (11.87 MJ) against the designed capacity of 10 MJ. Addition of fins reduces the effective storage volume of concrete; however, the reduction in storage material volume is not significant as compared to decrement in discharging time. Increasing the heat transfer fluid flow rate beyond 0.25 m/s is not significant for all three SHS modules.

Acknowledgements The authors sincerely thank the Department of Science and Technology (DST), Government of India, for the financial support (Project No: DST/TM/SERI/2K10/53(G)).

References

1. Z. Yang, S.V. Garimella, Thermal analysis of solar thermal energy storage in a molten salt thermocline. Sol. Energy **84**, 974–985 (2010)
2. S. Khare, C. Knight, S. McGarry, Selection of materials for high temperature sensible energy storage. Sol. Energy Mater. Sol. Cells **115**, 114–122 (2013)
3. A. Fernandez, M. Martinez, M. Segarra, I. Martorel, F. Cabeza, Selection of materials with potential in sensible thermal energy storage. Sol. Energy Mater. Sol. Cells **94**, 1723–1729 (2010)
4. A. Gil, M. Medrano, F. Cabeza, State of the art on high temperature thermal energy storage for power generation, Part 1—concepts, materials and modellization. Renew. Sustain. Energy Rev. **14**, 31–55 (2010)
5. R. Tamme, D. Laing, W.D. Steinmann, Advanced thermal energy storage technology for parabolic trough. J. Sol. Energy Eng. **126**, 794–800 (2004)
6. D. Laing, W.D. Steinmann, R. Tamme, C. Richter, Solid media thermal storage for parabolic trough power plants. Sol. Energy **80**, 1283–1289 (2006)
7. D. Laing, C. Bahl, T. Bauer, D. Lehmann, Thermal energy storage for direct steam generation. Sol. Energy **85**, 627–633 (2011)
8. E. John, W.M. Hale, R.P. Selvam, Development of a high-performance concrete to store thermal energy for concentrating solar power plants, in *Proceedings of the ASME 5th ICES Washington*, DC, USA (2011)
9. F. Agyenim, P. Eames, M. Smyth, Heat transfer enhancement in medium temperature thermal energy storage system using a multi tube heat transfer array. Renew. Energy **35**, 198–207 (2012)
10. B.R. Nandi, S. Bandyopadhyay, R. Banerjee, Analysis of high temperature thermal energy storage for solar power plant, in *IEEE Third International Conference on Sustainable Energy Technologies*, Nepal (2012)
11. D. Sragovich, Transient analysis for designing and predicting operational performance of a high temperature sensible thermal energy storage system. Sol. Energy **43**, 7–16 (1989)
12. L. Miro, M.E. Navarro, P. Suresh, A. Gil, A.I. Fernandez, L.F. Cabeza, Experimental characterization of a solid industrial by product as material for high temperature sensible thermal energy storage (TES). Appl. Energy **113**, 1261–1268 (2014)
13. R. Anderson, S. Shiri, H. Bindra, J.F. Morris, Experimental results and modelling of energy storage and recovery in a packed bed of alumina particles. Appl. Energy **119**, 521–529 (2014)
14. L. Prasad, P. Muthukumar, Design and optimization of lab-scale sensible heat storage prototype for solar thermal power plant application. Sol. Energy **97**, 217–229 (2013)

15. H. Niyas, L. Prasad, P. Muthukumar, Performance investigation of high-temperature sensible heat thermal energy storage system during charging and discharging cycles. Clean Technol. Environ. Policy **17**, 501–513 (2015)
16. M.Y. Haller, C. Cruickshank, W. Streicher, S.J. Harrison, E. Anderson, S. Furbo, Methods to determine stratification efficiency of thermal storage processes—review and theoretical comparison. Sol. Energy **83**, 1847–1860 (2009)
17. Y. Tian, C.Y. Zhao, A review of solar collectors and thermal energy storage in solar thermal applications. Appl. Energy **104**, 538–553 (2013)

Development of Phase Change Material-Based Temperature Regulation Facility Inside Protected Agriculture Structure

Priyabrata Santra, P.C. Pande, A.K. Singh, Surendra Poonia and D. Mishra

Abstract In arid agriculture, protected cultivation inside structures with polythene or shade net as covered material may be found beneficial to avoid aberrant weather conditions. However, significant amount of heat load is generated inside the structure due to high solar irradiation and trapping of outgoing longwave radiation due to protected covers, which lead to uncongenial conditions for sustainable plant growth. With a view to regulating temperature inside the protected cultivation structure, phase change material (PCM)-based temperature regulation facility was tried to develop through laboratory trials and experiments. Mixtures of paraffin liquid and paraffin wax in different ratios were prepared in the laboratory and heat change of the mixture material was recorded during heating and cooling process in a controlled water bath experimental setup. A mixture of paraffin liquid and wax in the ratio 10:1 was found suitable as a phase change material with melting temperature at about 35–38 °C. In the second stage, two heat exchanger structures consisting of GI sheet tank of 6.5 L capacity and aluminium fins attached at both sides of the tank for better heat transfer with air were designed and fabricated and these were tested inside small experimental structures with fibre glass sheet casing and covers. A decrease of inside temperature by 8–10 °C has been observed inside prototype structure with the incorporation of PCM material compared to one without it. Subsequently, the system was kept inside the PV clad enclosure (13.5 m^2) and the performance of the enclosure was studied with chilli crop grown inside. The amount of heat stored by a single PCM storage system with 2.5 kg PCM materials was found about 1100–1200 kJ. An estimate indicates that it is sufficient to reduce the inside temperature by 1 °C for four hours if number of air change rate is 30 per hour. However, long-term trials are required to ascertain number of cycles of heating and cooling of the material.

P. Santra (✉) · P.C. Pande · A.K. Singh · S. Poonia · D. Mishra
Division of Agricultural Engineering for Arid Production Systems,
ICAR-Central Arid Zone Research Institute, Jodhpur 342003, Rajasthan, India
e-mail: Priyabrata.Santra@icar.gov.in

S. Poonia
e-mail: poonia.surendra@gmail.com

L. Chandra and A. Dixit (eds.), *Concentrated Solar Thermal Energy Technologies*, Springer Proceedings in Energy, https://doi.org/10.1007/978-981-10-4576-9_17

Keywords Protected agriculture · Arid and semiarid region · Phase change material, paraffin wax

1 Introduction

Cultivation of crops in greenhouse or protected structure has been increased tremendously in recent times and also has been spread from temperate regions to the warmer regions of tropics and subtropics. Plentiful technologies have been developed for heating the inside environment of greenhouse for optimum plant growth [1]. However, in hot arid areas, cultivation in greenhouse is very difficult as the temperature reaches to a very high level to sustain plant growth [2, 3]. Under such situation requirement for cooling the greenhouse microclimate environment is the necessity, but it is very energy and water intensive. Hence, in such regions, reduction of air temperature inside the greenhouse or the regulation of temperature closer to the ambient temperature during summer is necessary for successful crop production. A typical diurnal variation of temperature and global solar radiation during late winter months inside a shade net structure has been plotted in Fig. 1. It has been found that global solar radiation has been decreased to almost one-third of the radiation received outside by using 75% shade net cover. The temperature inside structure during morning time has also been found higher than outside temperature and in summer such increase is expected to be much higher.

Therefore, cooling is considered as the basic necessity for greenhouse crop production in tropical and subtropical regions to overcome the problems of high temperatures during summer months. Development of suitable cooling system that provides congenial microclimate for crop growth is a difficult task as the design is closely related to the local environmental conditions. Different cooling techniques for greenhouse crop cultivation may be found in Kumar et al. [4]. Fan pad cooling system has been widely used for greenhouse cultivation however, requires large amount of water during evaporation-based cooling. Since water resources are becoming scarce in future such fan-pad cooling system may not be viable in future specifically in hot arid regions.

Energy storage plays important roles in regulating the heat load inside a protected structure. Solar energy is available only during the day, and hence, efficient thermal energy storage system may help in eliminating the excess heat generated during sunshine hours. The most commonly used method of thermal energy storage is the sensible heat method. Water is still used for heat storage in liquid-based systems, while a rock bed is used for air-based systems. The latent heat method of storage has attracted a large number of applications, because this method of heat energy storage provides much higher energy storage density with a smaller temperature swing when compared with the sensible heat storage [5]. Phase change material (PCM) may also be used as thermal storage of heat and thus to cool the greenhouse environment, however, reports are available only for heating purpose [6] and not for cooling. Shading of greenhouse structure sometimes found effective

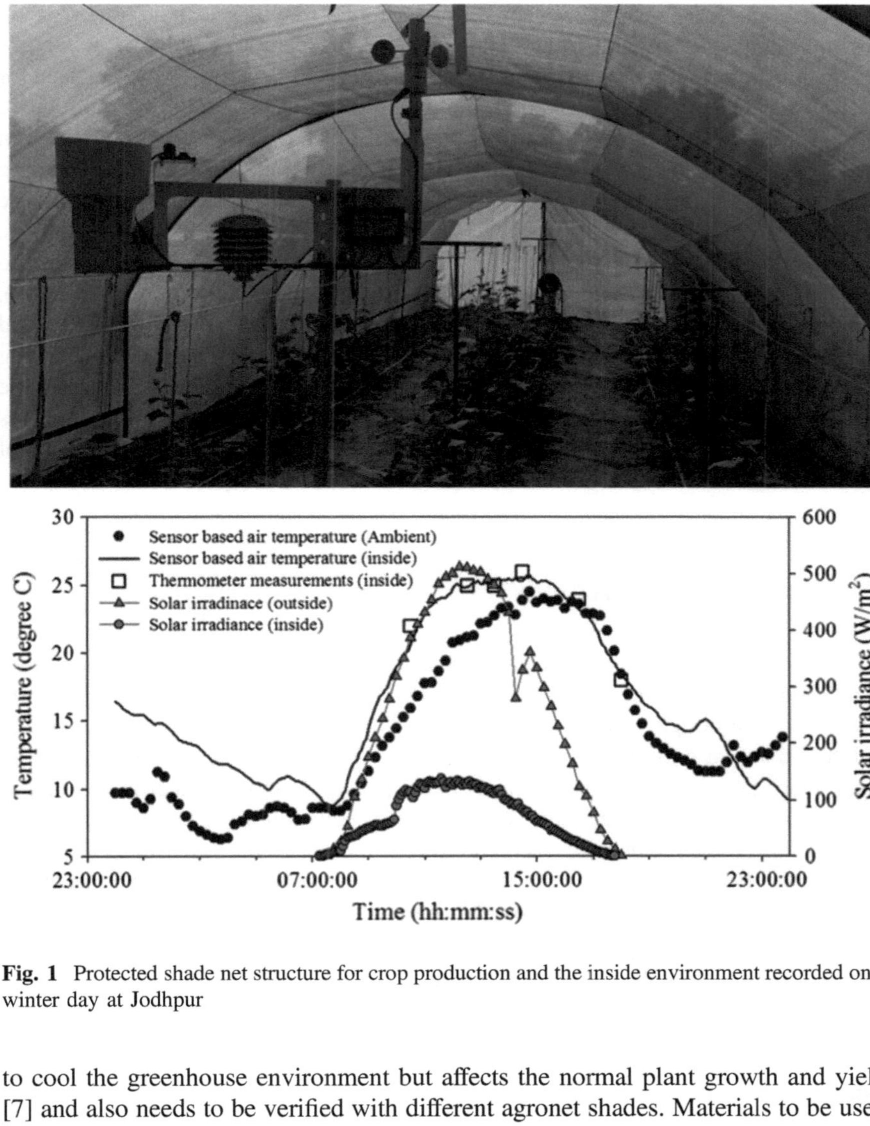

Fig. 1 Protected shade net structure for crop production and the inside environment recorded on a winter day at Jodhpur

to cool the greenhouse environment but affects the normal plant growth and yield [7] and also needs to be verified with different agronet shades. Materials to be used for phase change thermal (PCM) energy storage must have a large latent heat and high thermal conductivity. During the past 40 years most commonly used PCM materials are hydrated salts, paraffin waxes, fatty acids and eutectics of organic and non-organic compounds [8–10]. PCM has been successfully applied for cooling of buildings and even applied on textiles [11, 12]. Recently, the encapsulation of PCM has been successfully demonstrated for its wider application [13, 14].

Depending on the applications, the PCMs should first be selected based on their melting temperature lying in the practical range of operation, which is about 25–35 °C for protected agriculture system. Materials that melt below 15 °C are used for storing coolness in air conditioning applications, while materials that melt above 90 °C are used for absorption refrigeration. All other materials that melt between these two temperatures can be applied in solar heating and for heat load levelling applications. Commercial paraffin waxes are cheap with moderate thermal storage densities (~200 kJ/kg or 150 MJ/m^3) and a wide range of melting temperatures. They undergo negligible subcooling and are chemically inert and stable with no phase segregation. However, they have low thermal conductivity (~0.2 W/m °C), which limits their applications. Pure paraffin waxes are very expensive, and therefore, only technical grade paraffins can be used. Keeping in mind the cooling requirement of protected cultivation system and potential of paraffin wax as suitable PCM for heat storage, it was tried to develop a material mixture of different grades of paraffin waxes and a heat exchanger for regulating the inside temperature of protected agriculture system at hot arid region of India.

2 Materials and Methods

Mixtures of paraffin liquid and paraffin wax was prepared as PCM at laboratory with different ratios. For proper mixing of two materials, they were first heated and then mixed thoroughly using a stirring rod. Temperature changes of the prepared mixture material during heating and cooling process was determined using a water bath. During heating experiment, water in the bath and mixture material was heated from its normal room temperature and change of temperature of both water and the material was recorded using glass thermometer. In the cooling experiments, mixture materials were kept inside water bath, which was preheated to 75–80 °C. Apart ftom these, hot water bath experiment at a predefined temperature (75–80 °C) was also carried out using a thermostat system. Heat gain and heat loss during heating and cooling experiment was calculated by the following equation:

$$q = m.s.\Delta t \tag{1}$$

where

- q amount of heat loss or gain,
- m mass of water,
- s specific heat capacity of water, and
- Δt change in temperature.

After identifying a suitable mixture as PCM with melting temperature at around 35 °C, it was first tested inside a small prototype greenhouse structure and then in a PV clad enclosure. A heat exchanger of GI tank along with aluminium fins surrounding it was prepared and PCM was stored inside it. The PCM was filled inside the tank up to 80–85% level and then kept inside prototype structure. Temperature of both inside and outside of the prototype structure was monitored at half hourly interval. Prototype structure was prepared using transparent PVC sheet at top and surround wall, whereas the floor was prepared with plywood sheet. The performance of PCM was also tested inside a PV clad enclosure with chilli crop grown inside. Sensible heat storage and latent heat storage of the heat exchanger system was quantified using the Eq (1). The specific heat capacity of the heat exchanger material was considered as 0.22 kcal/kg °C for aluminium, 0.122 kcal/kg °C for GI and 0.24 kcal/kg °C for air.

3 Results and Discussion

3.1 Water Bath Experiment

Typical temperature profiles of PCM material mixture are presented in Fig. 2. It has been observed that during phase change, temperature of PCM material remained almost similar for a period, which was about 32–35 °C. It was also noted that when phase changed occurred, physical state of PCM material changed from semi-solid to semi-liquid state.

3.2 Constant Hot Water Bath Experiment

Similarly the temperature profile of PCM material during hot water bath experiment was presented in Fig. 3. Characteristically, it has been observed that temperature of PCM material mixture remained constant at about 34–35 °C after 2 min from the start of experiment.

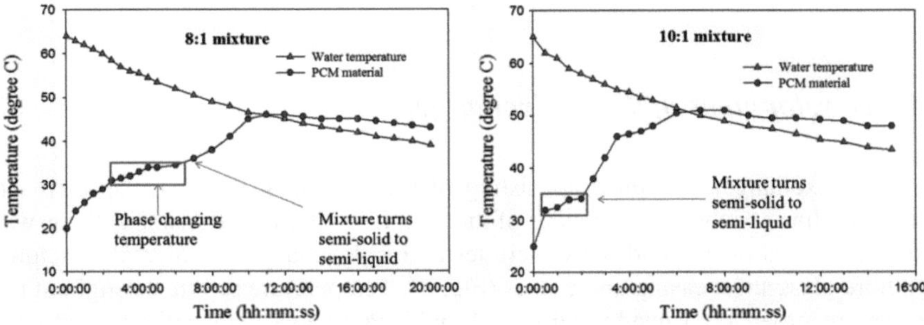

Fig. 2 Temperature profile of PCM material in water bath experiment with 8:1 and 10:1 mixture of paraffin liquid and wax

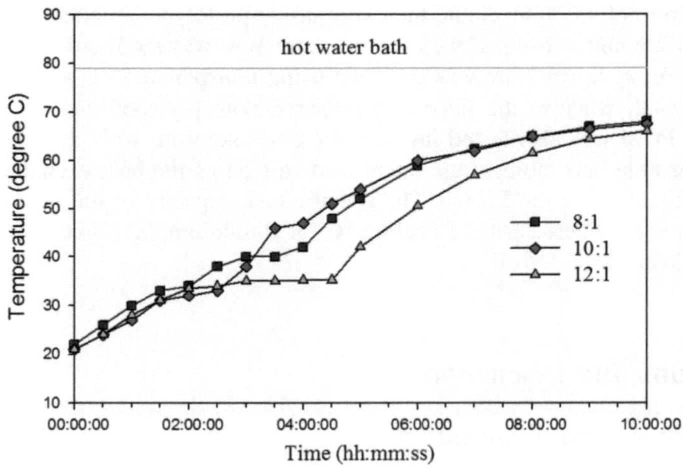

Fig. 3 Temperature profile of PCM material in hot water bath experiment with 8:1, 10:1 and 12:1 mixture of paraffin liquid and wax

Table 1 Phase changing temperature of PCM material prepared with mixing of paraffin liquid and wax at different ratios as obtained from water bath experiment

Mixture of paraffin liquid and wax	Density (g/cm^3)	Phase changing temperature
6:1	0.841	37–39
8:1	0.808	40–42
10:1	0.765	31–35
12:1	0.861	34–35

3.3 Phase Changing Temperature

Identified phase change temperature of different PCM material mixture is given in Table 1. Mixture of paraffin liquid and wax at a ratio of 10:1 was found best suitable for regulating temperature of protected agriculture structure at about 35 °C.

3.4 Development of Heat Exchanger

Two heat exchanger structures consisting of GI sheet tank of 6.5 L capacity and aluminium fins attached at both sides of the tank for better heat transfer with air was designed and fabricated and these were tested inside small experimental structures with fibre glass sheet casing and covers (Fig. 4). Temperature profile throughout the day was recorded both inside and outside with PCM load and without load condition. It was observed that temperature inside structure without any PCM load reached to 60–62 °C, when ambient was about 33–34 °C. With PCM load of 5.2 L,

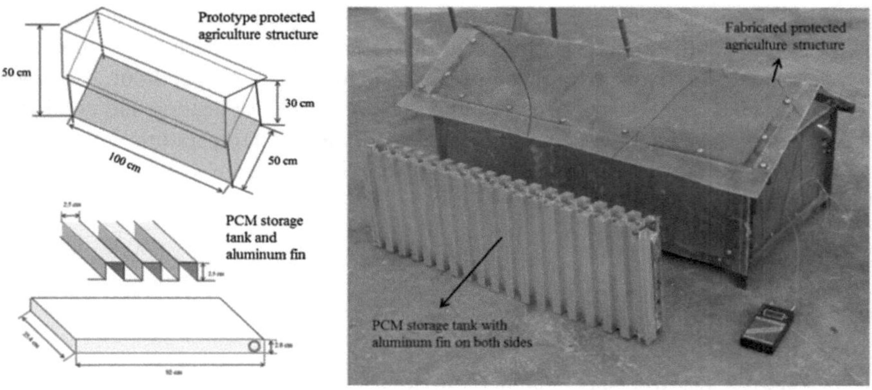

Fig. 4 Heat exchanger structure for storing PCM material-and prototype-protected agriculture structure for experiment

Fig. 5 Experimental trial of PCM material inside protected structure with chilli crop grown inside it

inside temperature was found 53–54 °C, when ambient was 33–34 °C. Thus, a reduction in inside temperature by 7–8 °C was observed by using PCM material.

3.5 Testing of PCM Material Inside Protected Structure

Subsequently, the system was kept inside the PV clad enclosure and the performance of the enclosure was studied for growing chilli (Fig. 5). The amount of heat stored by a single PCM storage system with 2.5 kg PCM was found about 1100–1200 kJ. An estimate indicates that it is sufficient to reduce the inside temperature by 1 °C for four hours if number of air change rate is 30 per hour. However, long-term trials are required to ascertain number of cycles of heating and cooling of the material. Results have so far indicated regulation of temperature within desirable limits leading to reasonable growth of chilli plants with good flowering and bearing fruits (Fig. 5).

4 Conclusion

A mixture of paraffin wax and paraffin liquid was developed as phase change material for its application in thermal regulation of inside environment of protected agriculture structure. Heating and cooling experiments of the developed mixture in laboratory showed characteristic properties of a PCM, however, there is further need for detailed characterization of the material using DSC calorimeter. Furthermore, a heat exchanger made of GI sheet was developed for storage of developed PCM material, which may be further improved. In this study, the developed PCM was tested inside a prototype structure, but needs further testing inside protected agriculture systems, e.g. polyhouse, shade nets etc. Such experiments should focus on optimization of the quantity of PCM material to be required per unit area of protected structure for optimum thermal regulation inside it.

References

1. V.P. Sethi, K. Sumathy, C. Lee, D.S. Pal, Thermal modeling of solar greenhouse microclimate control: a review on heating technologies. Sol. Energy **96**, 56–82 (2013)
2. Z. Pek, L. Hayles, The effect of daily temperature on truss flowering rate of ornamental crops. J. Sci. Food Agric. **84**(13), 1671–1674 (2004)
3. G. Sharan, T. Madhavan, Cropping in semi-arid northwestern India in greenhouse with ground coupling shading and natural ventilation for environmental control. Int. J. Serv. Learn. Agric. **5**(1), 148–169 (2010)
4. K.S. Kumar, K.N. Tiwari, M.K. Jha, Design and technology for greenhouse cooling in tropical and subtropical regions: a review. Energy Build. **41**, 1269–1275 (2009)
5. M.M. Farid, A.M. Khudhair, S.A.K. Razack, S. Al-Hallaj, A review on phase change energy storage: materials and applications. Energy Convers. Manag. **45**, 1597–1615 (2004)
6. F. Berroug, E.K. Lakhal, Omari, M.El. Faraji, H.El. Qarnia, Thermal performance of a greenhouse with a phase change material north wall. Energy Build. **43**, 3027–3035 (2011)
7. H. Harrou, L. Guilioni, L. Dufour, C. Dupraz, J. Wery, Microclimate under agrivoltaic systems: is crop growth rate affected in the partial shade of solar panels. Agric. For. Meteorol. **177**, 117–132 (2013)
8. A.M. Khudhair, M.M. Farid, A review on energy conservation in building applications with thermal storage by latent heat using phase change materials. Energy Convers. Manag. **45**, 263–275 (2004)

9. Y. Yuan, N. Zhang, W. Tao, X. Cao, Y. He, Fatty acids as phase change materials: a review. Renew. Sustain. Energy Rev. **29**, 482–498 (2014)
10. A. Sharma, V.V. Tyagi, C.R. Chen, D. Buddhi, Review on thermal energy storage with phase change materials and applications. Renew. Sustain. Energy Rev. **13**, 318–345 (2009)
11. M.J. Huang, P.C. Eames, B. Norton, Thermal regulation of building-integrated photovoltaics using phase change materials. Int. J. Heat Mass Transf. **47**(12), 2715–2733 (2004)
12. S. Mondal, Phase change materials for smart textiles–An overview. Appl. Therm. Eng. **28** (11), 1536–1550 (2008)
13. P.B. Salunkhe, P.S. Shembekar, A review on effect of phase change material encapsulation on the thermal performance of a system. Renew. Sustain. Energy Rev. **16**(8), 5603–5616 (2012)
14. M. Karkri, M. Lachheb, Z. Nógellová, B. Boh, B. Sumiga, M.A. AlMaadeed, A. Fethi, I. Krupa, Thermal properties of phase-change materials based on high-density polyethylene filled with micro-encapsulated paraffin wax for thermal energy storage. Energy Build. **88**, 144–152 (2015)

Part VI
Solar Cooling

Optimum Composition of Alternative Refrigerant Mixture for Direct Expansion Solar-Assisted Heat Pump Using ANN and GA

K. Vinoth Kumar, Lokesh Paradeshi, M. Srinivas and S. Jayaraj

Abstract In this paper, methodology for finding optimum composition of binary mixture (R290/R1270) for Solar-Assisted Heat Pump (Direct Expansion) is framed. Experiments are conducted on DX-SAHP by using R22 under the metrological condition of Calicut, India. Properties of binary mixture such as vapour pressure, liquid density for mass percentage of R1270 ranging from 0 to 50% is taken from REFPROP 11® software. Artificial Neural Network (ANN) is integrated with Genetic algorithm (GA) in MATLAB 7.7.0® to obtain optimum composition of binary mixture, where maximized EPR (Energy Performance Ratio) and minimized cost of refrigerant are two objective functions. The result shows that, the optimum compositions of binary mixture are 70% of R290 with 30% of R1270. Unique feature of this study is that, ANN integrated with GA technique can be used to find best optimum composition of alternative refrigerant mixture by mass percentage for any given system, provided experimental data of that system should be available for R22 or any other refrigerant.

Keywords Direct expansion solar-assisted heap pump · ANN · GA · Alternative refrigerant · Optimize mixture composition

1 Introduction

Heat pumps are one of the energy-efficient devices in the thermal applications. Integrating solar energy with heat pump technology is one of the best alternative technologies to improve the performance of the system performance. The solar-assisted heat pump system (Direct Expansion) performance is greatly influenced by the operating parameters such as ambient temperature, solar intensity and type of collector used in the system (acting as evaporator) and followed by wind

K. Vinoth Kumar (✉) · L. Paradeshi · M. Srinivas · S. Jayaraj
Department of Mechanical Engineering, National Institute of Technology, Calicut 673601, India
e-mail: vkvinoth3@gmail.com

L. Chandra and A. Dixit (eds.), *Concentrated Solar Thermal Energy Technologies*, Springer Proceedings in Energy, https://doi.org/10.1007/978-981-10-4576-9_18

velocity. The selection of working fluid also plays important role in the system performance. Because of good thermodynamic and thermosphysical properties of the R22, it has been widely used in the different applications. But the environmental properties such as Global warming potential (GWP) and Ozone depletion potential (ODP) for R22 are higher compared to hydrocarbons. As ODP and GWP value for R22 is high, Montreal and Kyoto protocol [1–3] made an agreement so as to phase out the usage of R22. So, there is a need to look for alternative refrigerant for DX-SAHP and also that should be easily available.

Numbers of investigations are reported for the replacement for R22 in different thermal applications. In that Hydrocarbons (HC) are one of the promising alternatives for the R22. The hydrocarbons are suffering from the flammability issue, hence its usage was stopped for many years for the safety concern. But considering environmental issue, restriction of usage of hydrocarbons is somewhat relaxed [4, 5]. Ki-Jung et al. [5, 6] experimentally studied the performance of hydrocarbon such as R432A and R433A in heat pump bench tester for residential air conditioner applications. Ki-Jung et al. [7] also investigated the heat pump performance charged with R170/R290 mixture. Purkayastha et al. [8] tested the performance of refrigeration and heat pump using the propane and LPG (Liquefied Petroleum Gas) mixture. In DX-SAHP, Abou et al. [9] tested with R134a, Mohanraj et al. [10] tested with R407C and Wang et al. [11] tested with R245fa.

From the literature survey, it is observed that no work has been carried out on DX-SAHP by using HCs. Among available HCs, R290 and R1270 have saturation pressure close to the compressor specification which is used in experimental setup. So in this paper, properties of R290 and R1270 are studied. Also it is observed that alternatives are chosen based on theoretical studies and its performance are observed by conducting experiment on it. No other method is performed so far so as to predict the performance of alternative refrigerant without conducting experiment on it. Similarly, there is no method implemented to find optimum composition of refrigerant mixture by considering cost of refrigerant and EPR of the system. The scope of this research article is to frame methodology so as to find best optimum composition of alternative refrigerant mixture. An ANN integrated with GA in MATLAB®7.7.0 has been developed to predict the optimum composition of refrigerant mixture. The experimental results from experimentation by using R22 and properties of refrigerants from REFPROP®11.0 were used in this method.

2 Experiment

2.1 Description of Experimental Setup

The present experimental setup mainly include the compressor (Rated speed and power of 1500 RPM and 1022 W respectively), an air-cooled condenser and thermostatic expansion valve and evaporator (Flat plate collector; unglazed type

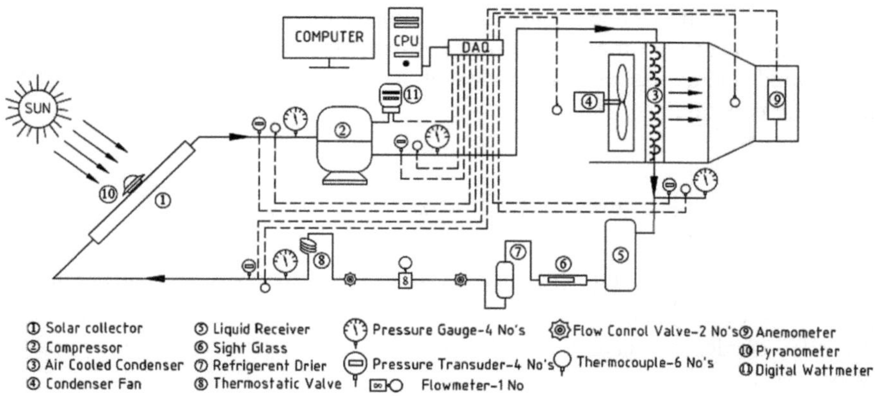

Fig. 1 Schematic diagram of the DX-SAHP system experimental setup

with 2 m^2 total area). The high pressure side of the system consists hermetically sealed compressor was connected to air-cooled condenser. The low-pressure side system consists expansion valve was connected to evaporator/solar collector. Heat pump circuit was completed by connecting the both circuits (high and low) as shown in Fig. 1. The evaporator/solar collector of the system have 0.8 mm copper absorber plate with 110 mm tube spacing. 25 mm-thick glass wool insulation was applied to the back surface of the collector to avoid the heat losses (both convention and radiation) from the collector. To obtain the maximum solar insolation on the collector face, it was tilted to at an angle 20° facing to south.

Pressure, temperature, flow rate and energy consumption were measured during the experimentation with the help of required instrumentation. Resistance temperature detector (RTDs) and pressure transduces were connected to each component, at inlet and out let of the system (Compressor, condenser, expansion valve and solar collector). Four mechanical pressure gauges were also installed to check any leakages in the system. Similarly the ambient temperature, inlet and outlet of the condenser air temperature are recorded with help of another set of RTDs. A turbine type flow metre was installed to measure the flow rate of the refrigerant in the system. Single-phase digital energy metre was used to note down the power consumption of the compressor. All the instruments were calibrated before using and accuracy limits are within the range. The experiments were carried out during February month of 2015 between 8 am and 5 pm at the Solar Energy Centre of National Institute of Technology, Calicut (latitude of 11.15°N, longitude of 75.49°E).

2.2 Uncertainty Analysis

The experiments were associated uncertainties; this is due to mainly measurement of various parameters such as the pressure, temperature and flow rate at various salient points in the system circuit as shown in Fig. 1. The uncertainty presented in the measurement of compressor power consumption, heating capacity of the condenser and energy performance ratios of the system were calculated using the following equation:

$$u_f = \left[\left(\frac{\partial f}{\partial x_1} u_{x1} \right)^2 + \left(\frac{\partial f}{\partial x_2} u_{x2} \right)^2 + \ldots\ldots + \left(\frac{\partial f}{\partial x_n} u_{xn} \right)^2 \right]^{1/2} \quad (1)$$

The uncertainties presented in the experimentation of DX-SAHP system for the parameters were calculated and found to be for compressor power consumption ±1.14%, condenser heating ±2.42% and for energy performance ratio ±2.8% respectively.

3 Study on R290 and R1270

Thermodynamic properties, thermophysical properties, chemical properties and environmental properties are to be studied when we want to choose an alternative refrigerant.

3.1 Thermodynamic Properties

Vapour saturation pressure for R290 and R1270 compare to R22 for temperature ranging from 0 to 80 °C are 10.25% lower and 8.27% higher respectively. Compressor which we use in experimental setup is meant for R22. According to compressor specification, vapour saturation pressure of alternative refrigerant should not be more than 5% of R22. Thus using R1270 alone in setup is not safe, whereas R290 alone can be used. As R290 have saturation pressure lower than R22, it may reduce system performance. So to increase pressure some percentage mass of R1270 can be mixed with R290. Compressor pressure ratio for R290 and R1270 compare to R22 at constant condenser temperature of 70 °C for temperature ranging from 0 to 80 °C are 7.34% lower and 10.23% lower respectively. Lower the compressor ratio lower will be the compressor discharge temperature. It is desirable to use refrigerant having low compressor pressure ratio, concerning the safety of compressor reed valve.

Liquid density for R290 and R1270 when compared to R22 for temperature ranging from 0 to 80 °C are 58.54% lower and 57.79% lower respectively. Quantity of refrigerant to be charged depends on their density. As lower the density, lower will be the requirement of refrigerant to be charged, it is desirable to use R290 and R1270 as alternative. Latent heat for R290 and R1270 when compared to R22 for temperature ranging from 0 to 80 °C are 84.2% higher and 81.21% higher

Table 1 Average percentage variation of R290 and R1270 as compared to R22

S. No.	Properties	R290	R1270
1	Vapour pressure	10.25 ↓	08.27 ↓
2	Compressor pressure ratio	07.34 ↓	10.23 ↓
3	Liquid density	58.54 ↓	57.79 ↓
4	Latent heat	84.2 ↑	81.21 ↑

respectively. A high latent heat is accompanied by a low specific volume in the vapour state. This decreases the power consumption and also reduces the required compressor displacement. So it is advantage to go for R290 and R1270. The thermodynamic properties of R290, R1270 and R22 for given operating temperature range are compared and listed in Table 1.

3.2 Thermo Physical Properties

Liquid thermal conductivity for R1270 when compared to R22 for temperature ranging from 0 to 80 °C are 14.19% higher and 36.67% higher respectively. Vapour thermal conductivity for R1270 when compared to R22 for temperature ranging from 0 to 80 °C are 60.05% higher and 70.02% higher respectively. Hence, if we use R290 and R1270 higher heat transfer can be expected in condenser and evaporator section.

3.3 Chemical and Environmental Properties

R290 and R1270 are flammable and non-toxic. But the hydrocarbons are suffering from the flammability issues. The flammability depends on many factors such as refrigerant quantity used in the system, volume of the space which is occupied leaked refrigerant, location of the system etc. All these factors will contribute to the degree of hazard. Because of small quantity of refrigerant required for this system, refrigerant leak will not be a major problem issue by the use of R290 and R1270. Hydrocarbons generally show good chemical stability within the system. R290 and R1270 are found to be non-reactive and chemically stable with compressor constituents and tube of the system for the different working range of temperatures. Also R290 and R1270 are compatible with mineral oils and synthetic lubricating oils which are used in compressor. Critical and environmental properties for R290, R1270 and R22 are compared in Table 2.

By comparing all these properties, we can assure that either R290 alone or R290 with R1270 are suitable for the system. It is preferable to go for mixture. But only question is that what percentage by mass of R290 and R1270 we are going to use. So, in order to find the optimum percentage, following methodology has been framed.

Table 2 Comparison of critical and environmental properties

S. No.	Refrigerant	Critical temperature (°C)	Critical pressure (MPa)	Molecular weight (kJ/kg mol)	Boiling point (°C)	ODP	GWP
1	R22	96.14	4.99	86.46	−40.81	0.05	1700
2	R290	96.74	4.25	44.09	−42.10	0	20
3	R1270	91.06	4.55	42.08	−47.70	0	20

4 Finding Percentage Composition by Using ANN and GA

4.1 Cost of Refrigerant

To evaluate the cost of refrigerant, we should know the density of R22 and R290 with 0–50% by mass of R1270 for the temperature ranging from 30 to 50 °C. Those density values are taken from REFPROP®11.0 software. In case of binary mixture we should consider liquid phase density. As per the system specification, the maximum mass of R22 that is possible to fill the system is 1.1 kg. Using all those collected data sets we can find mass of the binary mixture. Mass of binary mixture can be evaluated by using following relation:

$$m_{\text{mix}} = \left(\frac{\rho_{\text{mix}}}{\rho_{\text{R22}}}\right) * m_{\text{R22}} \tag{2}$$

where m_{mix} and ρ_{mix} are the mass and density of binary mixture. ρ_{R22} is the density of R22. In Bhoruka Gases Limited, Bangalore, India, the cost of R290 is 2000 per kg and cost of R1270 is 3500 per kg. Thus cost of binary mixture (R290/R1270) for the temperature ranging from 30 to 50 °C can be evaluated by

$$C = (C_{\text{R290}} * m_{\text{mix}} * \rho_{\text{R290}}) + (C_{\text{R1270}} * m_{\text{mix}} * \rho_{\text{R1270}}) \tag{3}$$

where C is the total cost of binary mixture, C_{R290} and C_{R1270} is the cost of R22 and R1270 respectively.

4.2 ANN (Artificial Neural Network)

ANN (Artificial neural network) is one of the tools in MATLAB®. It functions similar to human nervous system. The network type that we used here is feed-forward back propagation. This type of architecture networks mainly includes the three layers (input, hidden and output) and contains several neurons. Framed networks trained with the help of obtained experimental data sets known as the input and output data sets. It then fits itself so as to predict output for whatever input we give, provided it should be within the range of the input that we used for training. In this methodology we have used three ANN models. This ANN models is trained by using trainlm as training function and learngdm as adaption learning function. The input and hidden layer are connected with help of Logsig transfer function, similarly the hidden and output layer are connected with similar function. The numbers of hidden layers and neurons to be used are found out by using methodology provided in Fig. 2.

In ANN I model, input parameter is the temperature ranging from 30 to 50 °C and output parameter is saturation pressure for R22 and R290 with 0–50% by mass of R1270. 100 datasets for each refrigerant are taken from REFPROP® software is used for training. Number of hidden layer used is 10 and in each hidden layer 20 number of neuron are used. In ANN II model input parameter is condenser pressure, which we get by conducting experiment on R22 and output parameter is corresponding EPR. 100 experimental datasets are used for training. Here one hidden layer and 10 neurons are used. In ANN III model, input parameter is saturation pressure for R290 with 0–50% by mass of R1270 for each temperature ranging from 30 to 50 °C and output is cost of binary mixture. Here also one hidden layer and 10 neurons are used.

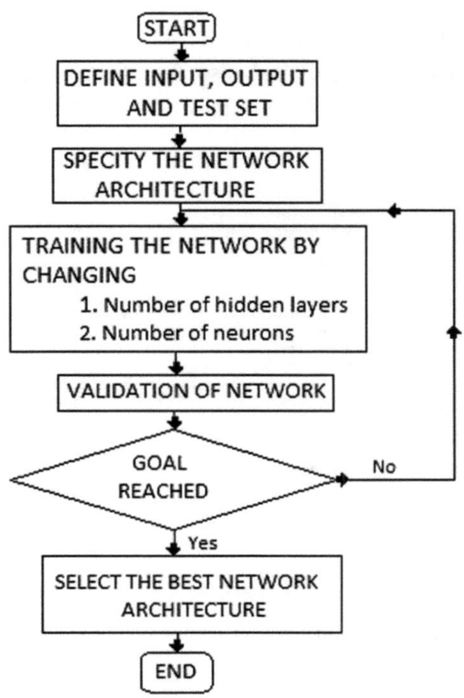

Fig. 2 Methodology for finding number of neurons and hidden layers in ANN

4.3 Genetic Algorithm

Genetic algorithm is also one of the tools in MATLAB®. It works based on the principle of biological evolution. It contains six fundamental issues and they are selection of function, chromosome representation, termination criteria, genetic operator making up the reproduction function, objective function and creation of the initial population. The objective function used here is

$$f = \frac{1}{(\text{Output from ANN III}) - (\text{Output from ANN II}) - x} \tag{4}$$

where x is the range and here fitness limit provided is zero.

4.4 Algorithm

The algorithm used for finding optimum mixture composition is shown in Fig. 3. Initially define saturation pressure for R22 and binary mixture taken from REFPROP® for temperature ranging from 30 to 50 °C and train ANN I. Then define condenser pressure and corresponding EPR from experimental data sets and train ANN II. Create a loop for temperature ranging from 30 to 50 °C with the increment of 2 °C. For 30 °C saturation pressure for different percentage by mass of R1270 is defined and cost of corresponding binary mixture is evaluated. Then ANN III is trained. And also for 30 °C, using ANN I saturation pressure for binary mixture for different percentage mass of R1270 is evaluated. Among the obtained

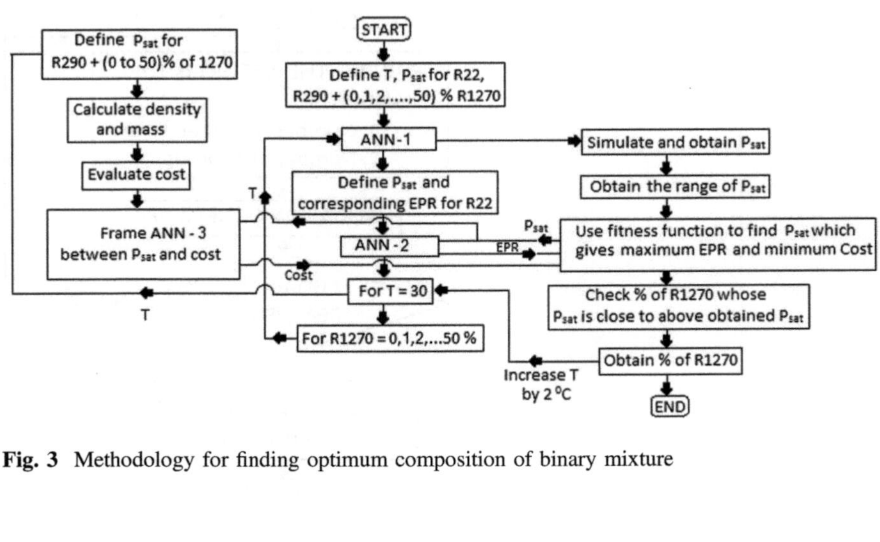

Fig. 3 Methodology for finding optimum composition of binary mixture

pressures maximum and minimum values have to be found out. This maximum and minimum values of pressure are provided to fitness function. Then the range of x is approximated based on the variation of EPR and cost with respect to the variation of % by mass of R1270. By using ANN II and ANN III, fitness function will evaluate EPR and cost for the provided range of pressure. As the result, fitness function will give pressure value for optimum value of EPR and cost. Then binary mixture for what percentage by mass of R1270 has pressure close to pressure obtained from fitness function is checked. And that will be the optimum % by mass of R1270 required for temperature 30 °C. Similarly run the loop by increasing temperature by 2 °C till 5 °C.

5 Result and Discussion

Table 3 shows the optimum percentage by mass of R1270 required to mix with R290 for the temperature ranging from 30 to 50 °C. Average percentage of R1270 obtained is 32.63%. Each time when we run the program average percentage of R1270 varies from 24 to 37%. So approximately we can go for 30% of R1270. For 30% of R1270 and 70% of R290 EPR can be predicted by using ANN II model. The predicted EPR is shown in Table 4. EPR value obtained from ANN II is close to value what we can expect when we run the setup by using 30 and 70% of R1270/R290. From this, we can confirm that the methodology framed to find the optimum percentage by mass of binary mixture is correct. Moreover, it takes less than 15 min to show the result. So it consumes very less time. As the vapour pressure of mixture and R22 are very close, the EPR will also be almost same.

Table 3 Optimum percentage by mass of R1270 and corresponding EPR predicted by program

Input	Output	
Temperature (°C)	% of R1270	EPR
30	15	1.43
32	17	1.43
34	15	1.44
36	19	1.47
38	23	1.53
40	35	1.78
42	47	2.04
44	49	2.08
46	48	2.23
48	49	2.47
50	42	2.53

Table 4 EPR for 30% of R1270 predicted by ANN II model

Input	Output	
Temperature (°C)	EPR	EPR of R22 (Expt)
30	1.16	1.17
32	1.17	1.22
34	1.19	1.25
36	1.25	1.30
38	1.37	1.47
40	1.62	1.92
42	1.93	2.10
44	2.04	2.15
46	2.09	2.20
48	2.29	2.41
50	2.49	2.68

6 Conclusion

The methodology of integrating ANN with GA is done successfully to predict the optimum percentage mass of binary mixture. The perfectness of this methodology is seen when we predict the EPR for optimum percentage composition of binary mixture. From this, it is observed that this methodology can be used to predict the optimum composition of binary mixture. More over this methodology can be applied to any system in any location in the world. It also consumes less time which is less than 15 min to run the program and reduces the need to conduct experiment to choose best alternative refrigerant mixture. Thus, it saves time as well as cost. As there is urgent need to choose alternative refrigerant considering environmental and energy issue, this methodology can be used.

References

1. United Nations Environmental Programme, *Montreal Protocol on Substances that Deplete the Ozone Layer, Final act, United Nations, New York* (1987)
2. Global Environmental Change Report, A brief analysis of the Kyoto protocol, 9(24) (1997)
3. Directive 2006/40/EC of the European Parliament and of the Council. Off. J. Eur. Union, 2006 (14.6.2006)
4. A. Cavallini, Working fluids for mechanical refrigeration. Int. J. Refrig. **19**, 485–496 (1996)
5. K.-J. Park, Y.-B. Shim, D. Jung, Performance of R433A for replacing HCFC22 used in residential air-conditioners and heat pumps. Appl. Energy **85**, 896–900 (2007)
6. K.-J. Park, Y.-B. Shim, D. Jung, Experimental performance of R432A to replace R22 in residential air-conditioners and heat pumps. Appl. Therm. Eng. **29**, 597–600 (2009)
7. K.-J. Park, D. Jung, Performance of heat pumps charged with R170/R290 mixture. Appl. Energy **86**, 2598–2603 (2009)

8. B. Purkayastha, P.K. Bansan, An experimental study on HC290 and a commercial liquefied petroleum gas (LPG) mix as suitable replacements for HCFC22. Int. J. Refrig. **21**, 3–17 (1997)
9. H.Z. Abou-Ziyan, M.F. Ahmed, M.N. Metwally, H.M.A. El-Hameed, Solar assisted R22 and R134a heat pump systems for low temperature applications. Energy Convers. Eng. Conf. **2**, 787–792 (1996)
10. M. Mohanraj, S. Jayaraj, C. Muraleedharan, A comparison of the performance of a direct expansion solar assisted heat pump working with R22 and a mixture of R407C liquefied petroleum gas. J. Power Energy **223**, 821 (2009)
11. X.D. Wang, L. Zhao, J.L. Wang, W.Z. Zhang, X.Z. Zhao, W. Wu, Performance evaluation of a low-temperature solar Rankine cycle system utilizing R245fa. Solar Energy **84**(3), 353–364 (2010)

Performance Evaluation of Solar–Biogas Hybrid Cold Storage System for Transient Storage of Horticultural Produce

Sampath Kumar Gundu, Asimkumar Joshi, Gokul Raj, Samir Vahora, Mukul Dubey and M. Shyam

Abstract Cold storage systems are predominantly used in production catchment area to minimize post harvesting losses for horticulture produce. Commercially available cold storage systems use huge amount of electricity and therefore are expensive. Renewable energy-based cold storage systems could be a clean, efficient, and cost-effective choice that can benefit farmers in rural areas, where electricity is scarce. We have developed a solar–biogas hybrid cold storage system that comprises of a vapour absorption machine (VAM), solar-evacuated tube collector (ETC) water heater, and solar photovoltaic module to meet the auxiliary power requirements. The system is integrated with all-brick masonry biogas plant of 50 m^3/day capacity to supplement the energy requirement during non-sunshine hours. Three cold chambers, each of 2–3 MT storage capacities have been set up. A comprehensive performance evaluation of the VAM and ETC with heat pipe was carried out. The maximum coefficient of performance (COP) of hot water driven VAM was found to be 0.71 and maximum efficiency of the ETC with heat pipe system was calculated to be 46%. This system has the potential to store 6–8 ton of horticultural produce for a period of 2–3 weeks at the desired temperatures range of 12–16 °C.

Keywords Cold storage · Solar photovoltaic-based refrigeration · Solar thermal refrigeration · Vapour absorption machine · Vapour compression refrigeration · Solar thermal · Evacuated tube collectors with heat pipe · Solar photovoltaic

---

S.K. Gundu (✉) · A. Joshi · G. Raj · S. Vahora · M. Dubey · M. Shyam
Sardar Patel Renewable Energy Research Institute,
Vallabh Vidhyanagar 388120, Gujarat, India
e-mail: sampath.bitsgoa@gmail.com

L. Chandra and A. Dixit (eds.), *Concentrated Solar Thermal Energy Technologies*,
Springer Proceedings in Energy, https://doi.org/10.1007/978-981-10-4576-9_19

1 Introduction

Cold storage plays an important role in preserving perishable items such as fruits and vegetables, preventing spoilage and protecting farmers from losses due to market rate fluctuations. Commercially available cold storage systems use electricity to produce the refrigeration effect which is very expensive [1]. Operation of such cold storage system is not possible in non-electrified regions. India receives abundant solar energy; average incident solar radiation varies from 4–7 kWh/day [1]. Solar energy-based cold storage system could be a promising alternative to conventional electricity-based systems that can benefit rural areas with no or irregular electricity supply.

Solar refrigeration is generally achieved in two ways (i) solar photovoltaic (PV) integrated refrigeration and (ii) solar thermal refrigeration [2]. In PV-integrated refrigeration system, electricity is produced directly by PV effect that runs the vapour compressor used in vapour compression refrigeration (VCR) system. In thermal refrigeration system, solar energy is converted into heat that drives the vapour absorption refrigeration (VAR) to produce the cooling effect. The solar–thermal type refrigeration system could be more efficient compared to PV-integrated refrigeration systems because high (>40%) sunlight to heat conversion efficiencies can be easily and cost effectively achieved compared to $\sim 10\%$ sunlight to electricity conversion efficiencies of the commonly marketed multi-crystalline silicon PV modules [2]. Additionally, solar PV-based cold storage system requires huge battery bank adding to the cost. A Solar thermal–biogas–hybrid cold storage system could be promising as the biogas source can supplement the energy requirement for the refrigeration system enabling its use during night hours, cloudy days, and rainy seasons without the requirement of expensive battery bank.

Single-effect vapour absorption type refrigeration using solar–thermal system has been reported by many authors. Rosiek and Batlles [3] investigated a single-effect solar absorption cooling system at the solar energy research centre, Spain. They found an average cooling capacity of 40 kW with an average coefficient of performance (COP) of 0.6. Ortiz et al. [4] performed an experiment with a Yazaki single-effect Li–Br–H_2O-fired chiller. They used flat-plate collectors of approximately 124 m^2 with vacuum tubular-collectors of 108 m^2 area. The system was capable to run within a temperature range of 70–95 °C with a capacity of 70 kW. Hidalgo et al. [5] designed an absorption cooling system for domestic cooling purpose where they used a flat-plate collector of 50 m^2 area with a single-effect absorption machine of 35 kW. The LiBr–H_2O working fluids showed a COP of 0.33 during the summer season in 2004. These systems however, used only solar thermal effect that can be harnessed during the day time or only when ample sunlight is available. We have developed a novel solar–biogas hybrid refrigeration system that harnesses sunlight during sunshine hours and uses auxiliary biogas source to fulfil energy requirement during non-sunshine hours, cloudy days, and rainy seasons. In this system, solar thermal ETC with heat pipe technology (medium temperature application) is integrated with the hot water driven Li–Br–water-based single-effect vapour absorption machine. A brick masonry biogas plant is integrated with the system for night or non-sunshine hour's operation.

2 Experimental Setup

The schematic of solar–biogas-based thermal refrigeration-based cold storage system shown in Fig. 1 was setup at the College of Food Processing Technology and Bio-energy, Anand Agricultural University (AAU), Anand, Gujarat.

The system comprises of a Li–Br–water-based vapour absorption machine (VAM) of 5 TR capacity developed by M/s Voltas Ltd. tailored to our requirements. The hot water temperature (85–90 °C) required to run the VAM was achieved through solar-evacuated tube collectors (ETC) with heat pipe during sunshine hours. Forty five modules of ETC with heat pipe each of area 3.27 m^2 was used to heat 1500 L water stored in hot water tank. During non-sunshine hours, a fixed dome all-brick masonry cattle dung (1250 kg/day) biogas plant of 50 m^3/day capacity was used to meet the additional energy requirements. The small electrical equipment such as pumps and fan coil units were powered by a 10 kW rooftop PV module procured from M/s Solex Energy Pvt. Ltd. The necessary controls and software were provided by M/s Mamata Energy Pvt. Ltd. The operational details of the system can be found in our previously published work [6]. The temperature data was collected using RTD type sensors and Datataker make datalogger. The data

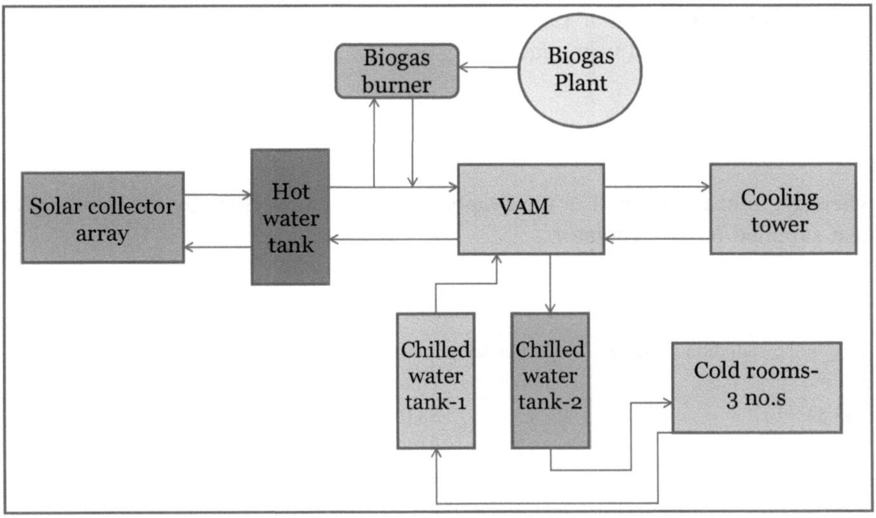

Fig. 1 Schematic of solar biogas refrigeration and storage system

was collected every second with average of 10 s being logged as a single data. Later, average of 15 min of stable data for three different days were taken as performance data of the particular subsystem for a particular radiation range for solar ETC field and a particular hot water inlet temperature range for VAM performance.

3 Results and Discussion

Performance of the solar thermal field was evaluated for varying average solar radiation. The efficiency of the ETC system was calculated using

$$\eta = \frac{mC_p(T_{c,o} - T_{c,i})}{A_c X G_t} \tag{1}$$

where m is the mass flow rate of the water in kg/s, C_p is the specific heat of water at constant pressure, $T_{c,o}$ is the outlet temperature of the water from the collector, $T_{c,i}$ is inlet temperature of the water to the collector, A_c is area of the collector, G_t is global radiation, η is efficiency of the collector.

The data shown in Table 1 clearly indicated that there was an increase in efficiency with increase in solar radiation. The maximum efficiency was found to be 46% at average solar radiation of 857.71 W/m^2 and water inlet temperature of 84.6 °C. Above 95 °C, steam generation was dominant which was not useful for driving the VAM; hence data for performance were taken only up to 95 °C outlet water temperature.

Further, performance of the VAM was evaluated for different hot water inlet temperatures varying in a range of 75–90 °C. The coefficient of performance (COP) was determined using equation

$$\mathrm{COP} = \frac{Q_e}{W_p + Q_g} \approx \frac{Q_e}{Q_g} \tag{2}$$

where Q_e is heat absorbed in the evaporator, W_p is work done by the pump, Q_g is heat supplied in the generator.

Table 1 Performance results of the solar thermal collector (5 rows, total area 147 m^2)

Avg. solar radiation (W/m^2)	Water temperature (°C)			Thermal efficiency (%)
	Initial	Final	Rise	
435.53	83.5	88.0	4.5	36
571.59	75.2	81.3	6.0	39
658.31	85.9	92.5	6.6	41
774.05	86.6	95.2	8.7	44
857.71	84.6	94.0	9.3	46

Table 2 Performance results of the VAM for different hot water inlet temperatures

Hot water temperature (°C)			Chilled water temperature (°C)			COP
Inlet	Outlet	Drop	Inlet	Outlet	Drop	
75.3	67.6	7.7	13.0	8.2	4.9	0.58
81.0	73.1	7.9	14.0	8.8	5.2	0.63
85.4	77.1	8.2	15.0	9.6	5.4	0.67
87.8	78.9	8.8	12.5	7.0	5.5	0.68
89.4	81.1	8.3	14.7	9.3	5.4	0.71

Table 3 Heating efficiency with biogas for heating of 1500 L water

S. No.	Burner gas consumed			Temperature rise		Efficiency (%)
	m^3	Period (h)	m^3/h	°C	°C/h	
Biogas (CV-5440 kcal/m^3)						
1	13.38	1.42	9.44	29.5	20.82	61
2	11.46	1.20	9.55	25.6	21.33	62
3	11.81	1.22	9.71	25.0	20.55	58
4	9.88	1.05	9.41	21.5	20.48	60
5	8.57	0.87	9.89	18.1	20.88	58
6	9.69	1.00	9.69	20.5	20.50	58
7	13.19	1.33	9.89	22.1	22.05	61
Average			9.65		20.95	60

The COP was found to vary between 0.58 and 0.71 (Table 2). The temperature of 3000 L water in the chilled water tanks dropped from 17 to 8 °C in 2.5–4.0 h depending upon the temperature of hot water at the inlet of the VAM. In the evening hours there was a steep drop in the solar radiation. The VAM, however, used the water available in the hot water storage tank and continued working for ~45 min after the solar hours.

The average heating efficiency of the gas burner for heating 1500 L of water using biogas was found to be 60%. Daily variation in efficiency is listed in Table 3. The maximum observed efficiency was 62% with an average biogas consumption of 9.65 m^3/h.

4 Conclusions

Maximum efficiency of the solar ETC heat pipe was calculated to be ~46% at 858 W/m^2 solar radiation and at hot water inlet temperature of 84.6 °C. The maximum COP of the VAM was found to be 0.71 at hot water inlet temperature of 89.4 °C. It was observed that minimum solar radiation of 500 W/m^2 is required to maintain the desired hot water temperature for VAM operation. Further, the storage temperature of 12–16 °C can be achieved for three cold rooms of 22.5 m^3 capacity each. Maximum and average heating efficiency of the burner with biogas was found to be 62 and 60% respectively. Further investigations on stored material quality, average power usage of solar and biogas, its techno-economic evaluation and comparison with conventional electrical cold storage systems are under progress.

Acknowledgements The present work was funded by "National Agricultural Science Fund (ICAR), Government of India". The authors are also thankful to Dr. B.S. Pathak (Ex-Director, SPRERI) for his valuable advice in carrying out this work. The authors are also thankful to Dr. V. Siva Reddy (Ex-Head, solar energy division, SPRERI), Tilak Chavda (Ex-Senior Scientist, SPRERI) for their valuable guidance for this work.

References

1. S.K. Maheshkumar, Non-conventional energy scenario and vision 2020 in India. ASM's Int. E-J. Ongoing Res. Manag. IT E-ISSN-2320-0065 (2015)
2. K.R. Ullah, R. Saidur, H.W. Ping, R.K. Akikur, N.H. Shuvo, A review of solar thermal refrigeration and cooling methods. Renew. Sustain. Energy Rev. Elsevier **24**, 499–513 (2013)
3. S. Rosiek, F.J. Batlles, Integration of the solar thermal energy in the construction: analysis of the solar-assisted air-conditioning system installed in CIESOL building. Renew. Energy Elsevier **34**(6), 1423–1431 (2009)
4. M. Ortiz, H. Barsun, H. He, P. Vorobieff, A. Mammoli, Modeling of a solar-assisted HVAC system with thermal storage. Energy Build. Elsevier **42**(4), 500–509 (2010)
5. M. Hidalgo, P.R. Aumente, M.I. Millán, A.L. Neumann, R.S. Mangual, Energy and carbon emission savings in Spanish housing air-conditioning using solar driven absorption system. Appl. Therm. Eng. Elsevier **28**(14–15), 1734–44 (2008)
6. V. Siva Reddy, K. Sampath, A. Gokul Raj, C. Tilak, Solar refrigeration technology for on-farm transient storage. Res. Article Cooling India **10**(7), 54–56 (2015)

Part VII
Process Heat

Design, Development and Performance Evolution of a Low-Cost Solar Dryer

Surendra Poonia, A.K. Singh, Priyabrata Santra and D. Mishra

Abstract Considerable amount of fruits and vegetables have been wasted due to non-availability of appropriate processing facilities in farmers field, in spite of high production in India with second rank in the world. A little of total fruit and vegetable productions in the country have been processed in industries. It has been noted that food crisis in most parts of the world is due to non-availability of processing and preserving facility rather than due to low production. Spoilage of fruits and vegetables can be reduced by drying and preserving them by suitable solar devices. Keeping in mind this requirement, a simple solar dryer has been fabricated at ICAR-CAZRI, Jodhpur. The dryer comprises a drying tray (0.75 m × 0.53 m) made of wooden frame and SS wire mesh and a glass sheet (4 mm thick) fitted on a wooden frame and a couple of hinges with opening and closing arrangements. Suitable openings for entry of ambient air and exit of water vapour have also been provided. The cost of the dryer is about Rs. 800. The drying tray serves dual purpose of solar collector and drying chamber. The maximum stagnation temperature observed inside the drying chamber was 63 °C on loading of 3 kg ber (*Ziziphus* sp.). It was 49 °C, when the outside ambient temperature was 27 °C on a clear sunny day (from 10 am to 4 pm) in the month of December–January. For drying of 3 kg ber fruits to a final weight of 1 kg using the developed dryer, it took about 7–8 days. It has been observed that the drying time for *Ziziphus* (ber) decreased by 50% in comparison to open sun drying. The developed dryer was also successfully tested to dry spinach (*Spinacia oleracea*), grated aonla (*Phyllanthus emblica*) and beetroot (*Beta vulgaris*). This solar dryer would be very useful for small and marginal farmers.

Keywords Low cost solar dryer · Solar collector · Drying chamber · *Ziziphus mauritiana* (ber)

S. Poonia (✉) · A.K. Singh · P. Santra · D. Mishra
Division of Agricultural Engineering for Arid Production Systems,
ICAR-Central Arid Zone Research Institute, Jodhpur 342003, Rajasthan, India
e-mail: poonia.surendra@gmail.com

P. Santra
e-mail: Priyabrata.Santra@icar.gov.in

L. Chandra and A. Dixit (eds.), *Concentrated Solar Thermal Energy Technologies*,
Springer Proceedings in Energy, https://doi.org/10.1007/978-981-10-4576-9_20

1 Introduction

India stands second in the world in producing fruits and vegetables, which are 12.6 and 14.0%, respectively. Nearly 76% of fruits and vegetables are consumed in fresh form of the total production and post-harvest losses are up to 20–22%. Only 2–4% of vegetable and fruit production are processed and preserved by industries. It was observed that food crisis in most of the countries is due to lack of ability to preserve food production rather than solely due to low production. In India, still 70% people are depending on agricultural practices and of this most people are not well off to afford advanced technologies and it is a major obstacle.

Drying produce in the open sun is still a practice in many countries. This method is very cheap but it creates a number of problems of contamination by foreign materials, which include dust, dirt, unwanted materials, infection due to insects and non-uniform drying. In the monsoon season, it becomes difficult to achieve optimum moisture content for safe storage and there are chances of materials also getting wet. India is endowed with abundant insolation. The arid western parts of India receive maximum irradiance, i.e. 7600–8000 MJ m^{-2} per annum, followed by semi-arid parts receiving 7200–7600 MJ m^{-2} per annum and least on hilly areas where solar irradiance is still appreciable, i.e. 6000 MJ m^{-2} per annum. Therefore, solar dryers can replace sun drying method. The produce dried in a solar dryer is not only economically viable but also of good quality and can return the cost incurred on solar dryer in one or two years. Therefore, small-sized solar dryers can be developed for cash crops and highly perishable products [1]. Some low-cost solar dryers were developed for entrepreneurs and marginal rural farmers as reported by Sharma et al. [2] and Fudholi et al. [3].

With this in view, ICAR-CAZRI, Jodhpur has developed a number of solar dryer of different designs during past three decades [4–10]. One major drawback behind wide-scale adoption of these previously developed solar dryers by farmers is the cost, which is not affordable by most of farming community. In order to make the dryer more economical, we have designed and developed a simple and small-sized solar dryer at ICAR-CAZRI, Jodhpur, India. The dryer that we have developed is a simple box-type solar dryer where the heat is provided by solar collector which also acts as drying chamber. Here the atmospheric air enters through inlet portion of the solar collector at the bottom end and the moist air gets exhausted through the outlet portion. The target stakeholders of the developed dryer are marginalized and poor farmers who cannot afford hi-tech facilities and equipment for preserving agricultural products, thus eliminating the unwanted and unpredictable food spoilage.

2 Design of Solar Dryer

The solar dryer is based on greenhouse effect wherein the solar irradiance enters into the chamber through top glass sheet which is trapped inside the drying chamber and thus increases the stagnation temperature. The experimental solar dryer

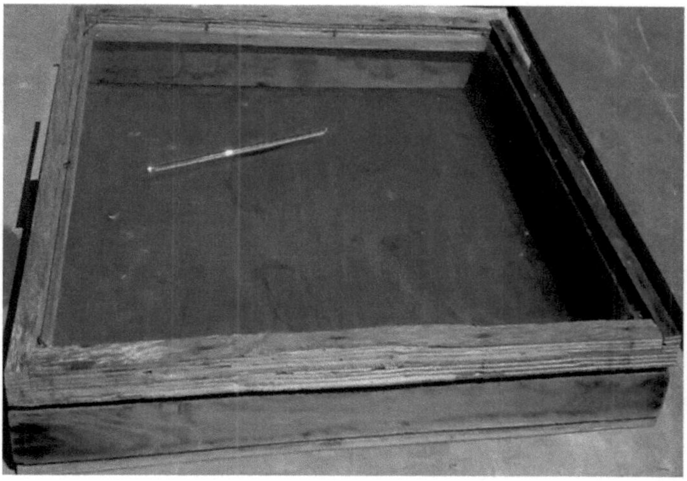

Fig. 1 A low-cost solar dryer developed at CAZRI, Jodhpur

comprises a drying tray (0.75 m × 0.53 m) made of wooden frame and SS wire mesh. A glass top made of clear window glass (4 mm thick) fitted on a wooden frame and a couple of hinges with opening and closing arrangements. Suitable openings for entry of ambient air and exit of water vapour have also been provided. The drying material can be put on the perforated SS wire mesh by opening the top wooden frame fitted with glass. The cost of the dryer is about Rs. 800. The drying tray serves the dual purpose of solar heat collector and drying chamber (Fig. 1).

3 Experimental Details and Results

In this dryer, the drying of *Ziziphus* (ber) was carried out during clear sky condition as on month of January 2016. About 3 kg of fresh ber was put inside the dryer with 80% moisture content on wet basis. Hourly values of total solar irradiance on the horizontal surface were measured by pyranometer coupled with integrator. The air temperature inside the dryer and ambient temperature were measured hourly from 10.00 to 17.00 h during the drying trials. The detailed results of the drying experiment of *Ziziphus* (ber) are given in Table 1.

From Table 1, it is clear that during the drying trials, maximum stagnation temperature observed inside the drying chamber was 63 °C and on loading 3 kg of ber the maximum temperature reduced to 49 °C, when the outside ambient temperature was 27 °C on a clear sunny day (from 10.00 to 17.00 h) in the month of January 2016. The moisture content of the ber decreased from 85 to 15% in 6 days and the rate of drying decreased rapidly for moisture content below 65%.

Table 1 Details of drying trials of *Ziziphus* (ber) in low-cost solar dryer

Date January 2016	Radiation on horizontal plane (kW/m^2)	Maximum temperature (in no load condition)		Radiation on horizontal plane (kW/m^2)	Maximum temperature (ber loaded condition)	
		Ambient (°C)	Inside dryer (°C)		Ambient (°C)	Inside dryer (°C)
10.00 h	680	20	36	640	21	34
11.00 h	820	23	58	880	23	40
12.00 h	920	25	62	980	25	44
13.00 h	980	27	63	960	27	49
14.00 h	900	26	59	920	26	47
15.00 h	720	26	57	640	26	44
16.00 h	490	25	49	390	24	40
17.00 h	210	22	40	220	25	35
Total/average	5720	24	53	5630	25	42

The efficiency of solar dryer (ratio of evaporative moisture to the solar irradiance on horizontal surface) is given by the relation given below:

$$\eta = \frac{ML}{A\int_0^\theta H_T \mathrm{d}\theta},$$

where

A Absorber area, m^2
H_T Solar radiation on horizontal plane, $J\ m^{-2}\ h^{-1}$
L Latent heat of vaporisation, $J\ kg^{-1}$
M Mass of moisture removed from the product, kg
θ Test duration, h
η Efficiency of the solar dryer.

The efficiency of the developed solar dryer was found about 16%. The developed dryer will be very useful for the people of rural areas. As there is an abundance of solar energy in our country, the rural people can use solar dryer for drying fruits and vegetables when this is a glut in the market and earn money by selling them.

3.1 Cost

The developed solar dryer comprises wooden flank, glass sheet (4 mm thick), and wire mesh. These materials are easily available and a village carpenter can fabricate this solar dryer. The cost of solar dryer is as low as Rs. 800/-.

4 Conclusion

The developed solar dryer could be very useful at the domestic level. Due to its low cost (Rs. 800/unit), it is very affordable. It can dry vegetable and fruits which can be packaged for use in off season. It can be used to dry fruits such as ber, aonla, beat root, leafy vegetables, spinach, coriander, chilies, fenugreek and other vegetables such as onion, garlic, carrot, etc. The use of such dryers at domestic level can go a long way in reducing postharvest losses.

References

1. P. Purohit, A. Kumar, T.C. Kandpal, Solar drying vs. open sun drying: a framework for financial evaluation. Sol. Energy **80**, 1568–1579 (2006)
2. A. Sharma, C.R. Chen, N. Vu Lan, Solar-energy drying systems: a review. Renew. Sustain. Energy Rev. **13**, 1185–1210 (2009)
3. A. Fudholi, K. Sopian, M.H. Ruslan, M.A. Alghoul, M.Y. Sulaiman, Review of solar dryers for agricultural and marine products. Renew. Sustain. Energy Rev. **14**, 1–30 (2010)
4. H.P. Garg, A. Krishnan, Solar drying of chillies in a solar cabinet dryer. Ann. Arid Zone **13** (4), 285–292 (1974)
5. H.P. Garg, K.P. Thanvi, Year round performance studies on a solar cabinet dryer at Jodhpur, in *Proceedings of 7th Meeting of All India Solar energy working group and Conference on the utilization of solar energy*, Ludhiana, pp. 142–145 (1975)
6. P.C. Pande, Performance studies on an improvised solar cabinet dryer, in *Proceedings of National Solar Energy Convention*, India. Allied Pub., India, pp. 1–5 (1980)
7. P.C. Pande, K.P. Thanvi, A solar dryer for maximum energy capture, in *Proceedings of National Solar Energy Convention*, IIT Delhi, India, pp. 4.009–4.0012 (1982)
8. P.C. Pande, K.P. Thanvi, Design and development of solar dryer cum water heater. Energy Convers. Manag. **31**(5), 419–424 (1991)
9. K.P. Thanvi, P.C. Pande, Development of a low cost solar agricultural dryer for arid regions of India. Energy Agric. **6**, 35–40 (1987)
10. K.P. Thanvi, P.C. Pande, Performance evaluation of solar dryers developed at CAZRI, Jodhpur, in *Solar Drying*, ed. by A.N. Mathur, Y. Ali, R.C. Maheshwari (Himanshu Publication, Udaipur, 1989), pp. 159–164

Performance Evaluation of Double Effect Solar Still

Chinmay Dave, Subarna Maiti, Bhupendra Kumar Markam, Pankaj Patel and Jatin Patel

Abstract Development of different useful decentralised solar gadgets is possible due to the abundant solar thermal radiation available particularly in the western part of India. Desalination is one technique to use this energy to produce distilled water using solar stills. More so, even highly saline water can be treated using solar stills. The output of conventional basin type solar still is only 1–2 LPD. In this, the latent heat of condensation is transferred to the surrounding and wasted. To use the latent heat and enhance the output a double effect type of solar still with two basins is fabricated in this work. A cotton cloth is kept on top of the lower basin to make a thin-film layer of continuously flowing water. Hot water is supplied to the lower basin through solar flat plate collector. It is seen that with increasing salt concentration in feed water results in a reduction of product output from a conventional still, but, in the modified double effect still, the product output remains the same.

Keywords Solar desalination · Double effect still · Continuous flow

C. Dave (✉) · J. Patel
Pandit Deendayal Petroleum University, Raisan, Gandhinagar 382007, Gujarat, India
e-mail: chinmaydave12@gmail.com

J. Patel
e-mail: jrpmech@gmail.com

S. Maiti · B.K. Markam (✉) · P. Patel (✉)
CSIR-Central Salt & Marine Chemicals Research Institute, Gijubhai Badheka Marg, Bhavnagar 364002, Gujarat, India
e-mail: bkmarkam@csmcri.res.in

P. Patel
e-mail: papatel@csmcri.res.in

S. Maiti
e-mail: smaiti@csmcri.res.in

L. Chandra and A. Dixit (eds.), *Concentrated Solar Thermal Energy Technologies*,
Springer Proceedings in Energy, https://doi.org/10.1007/978-981-10-4576-9_21

1 Introduction

There is plenty of water on the earth, but most of it is as seawater or as icebergs in the polar regions. Out of total water, 97% water is salty, other 2.6% is fresh water, but less than 1% is within human reach [3]. This shortage of drinking water is one of the biggest problems of the world. Purification of saline water is necessary which can be done by reverse osmosis, electrodialysis, multi-effect distillation and multi-stage flash, these methods either use fossil fuels or use electrical energy. Solar desalination is a technique with which we get distilled water with non-conventional solar energy.

There are several methods for the desalination of salty water using the solar energy, from that conventional solar still is one of the simplest methods to get the distilled water from saline water. In this method, sun rays pass through a glass top which also acts as a condensing surface, and are absorbed by blackened bottom of the basin. Water in the basin is heated which results in the generation of vapour. Gradually vapour in the basin increases resulting in more vapour pressure. The produced vapour goes up to the condensing surface due to the temperature difference between condensing surface and basin water surface. The vapour on the condensing surface is condensed and travels to the bottom due to gravity force, and is collected in a container, which is the distilled water required.

Conventional solar still is used from the ages to get the pure water, but it is having less product output per m^2 for a day. Maiti et al. [4] modified the design of solar still and made a stepped type solar still. It was having steps in the basin so as to increase the heat transfer area of the water resulting in more output of the solar still as compared to the conventional solar still. In this type of solar still due to steps, the area required for the product output is reduced.

Water vapour condenses on the glass surface during which it rejects heat to the surroundings, which is wasted. Rajaseenivasan and Murugavel [5] used this heat to enhance the temperature of water. They made a double basin solar still and experimentally found that product yield was 85% more in this still than the conventional solar still. Kalbasi and Esfahani [2] studied the effect of salt concentration on the solar still. They found out that with an increase in salinity the product output decreases.

El-Agouz et al. [1] studied the effect of the continuous flow of water on the solar still with and without makeup water. They compared three models, having an open loop for saline water, having closed loop for saline water without makeup water and last with makeup water. They concluded that with using the makeup water for a closed loop, the output we get is maximum.

2 Experimental Work

2.1 Setup

The experimental setup mainly consists of a flat plate collector and a double basin solar still. Flat plate collector is made of two different diameter aluminium pipes, 4.7 and 2.1 cm, having connected with a thin aluminium sheet to use the conduction heat transfer for increasing the temperature water. It is blackened to absorb more insolation from the sun. It is kept in a wooden box having a glass cover. The purpose of the collector is to provide heated water supply to the lower basin. The capacity of the collector is 4.5 L. The box is kept on a metal frame tilted at 21°. A ball valve is at the lower end of the flat plate collector. It regulates the water flow, coming from flat plate collector, in the lower basin of the solar still. Flat plate collector is charged from a 100 L capacity tank kept at a higher head.

Solar still has a surface area of 1.01 m^2. It has two basins, the lower basin, 119 cm × 85 cm × 2.5 cm, made of aluminium having 21 steps to increase the surface area for heat transfer and the product output of the basin. It is blackened to absorb more solar insolation. The lower basin is covered with a glass to prevent the water vapour from escaping and to reduce the thermal losses. It is having a hole in the second last pond for the case of overflow of water. The upper basin has perforated pipe for continuous flow of water droplets in the basin. To equally spread the water on the glass a white cotton cloth is used on the glass of the lower basin. The upper basin is having a 5 cm wooden structure covered with a glass. Product line made up of PVC is kept on the lower end of both the basins. A gap is kept between the lower basin and the upper basin structure at the lower end for removal of continuously flowing water. The still is kept on a 21° tilted metal frame having less height than collector frame. From the tank water continuously flows in the upper basin pipe. It is regulated by a globe valve.

Reflectors, made of aluminium sheets, having dimensions of 1.5 m × 1.5 m, is attached to a PVC sheet and fixed to the North–South of the system. The angle of the reflectors could be changed according to the seasonal movement of the sun. The unit is oriented facing south (Fig. 1).

2.2 Experimental Procedure

Heated water from the flat plate collector is charged to the lower basin at 11 o'clock every day till water comes from the overflow port. Overflowing of water stops after some time; then the ball valve is opened again and water is charged continuously in the form of droplets in the lower basin (Fig. 2).

In the upper basin water is continuously flowing at a very low flow rate 24 h. The water coming out from the lower end of the upper basin is collected in a

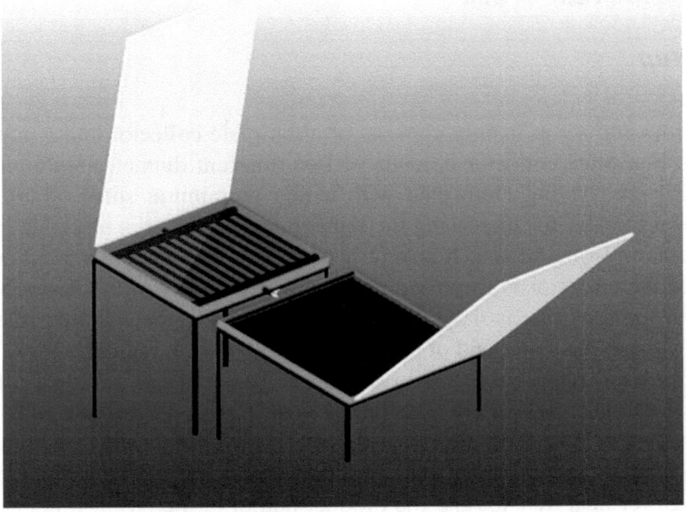

Fig. 1 Pictorial representation of the system at CSIR-CSMCRI, Bhavnagar

Fig. 2 System at CSIR-CSMCRI, Bhavnagar

container and is reused for the circulation. Flow in the upper basin is controlled by the globe valve.

During the experiment values of parameters like insolation on glass, Horizontal insolation, Upper glass temperature, Lower basin temperature at two points, Lower basin inlet temperature, Upper basin inlet–outlet temperature, Lower basin overflow temperature, Ambient temperature, Wind speed and Total dissolved solids are measured on the hourly basis from 11 o'clock to 5 o'clock. Night output is measured at 9:30 a.m. Again at 10 o'clock parameters are measured.

During the experiment flow rate of the water in the upper basin is varied and also the salinity in saline water is varied. The result of both the variation on the production of water is observed.

2.3 Measurements

Total dissolved solid is measured by Eutech CON 700 and Milwaukee SM801 pH/EC/TDS combined metre. Glass temperature is measured by an Infrared Thermometer. Ambient air temperature and wind speed is measured by the weather station. Remaining temperatures are measured using K-type thermocouple by HTC DT-305 contact thermometer. Solar insolation is measured with a Hukseflux LP02 pyranometer.

3 Result and Discussion

3.1 Effect of North–South Reflectors

To reduce the space required for the still, reflectors are used. Reflectors increase the solar radiation incident on the still which utilises the vertical space for still. We can get more heat for the same area of the still which directly affects the output of the still. In this system as the cotton cloth is used, it reduced the amount of solar radiation reaching the lower basin but using reflectors we can give more heat to the lower basin. In the figure variation of insolation due to reflectors is shown. It is evident that inclined solar flux is more than the horizontal solar flux due to reflectors (Fig. 3).

3.2 Effect of Salinity on Production

For finding the effect salt concentration of saline water produces, experiments are done. Saline waters with TDS of 1000, 10,000 and 20,000 are used. In the below figure the production of still with various saline waters is shown. It was seen that the output of the still remains similar in all the three conditions which is not the case for the normal solar still. In normal solar still with an increase in salinity product

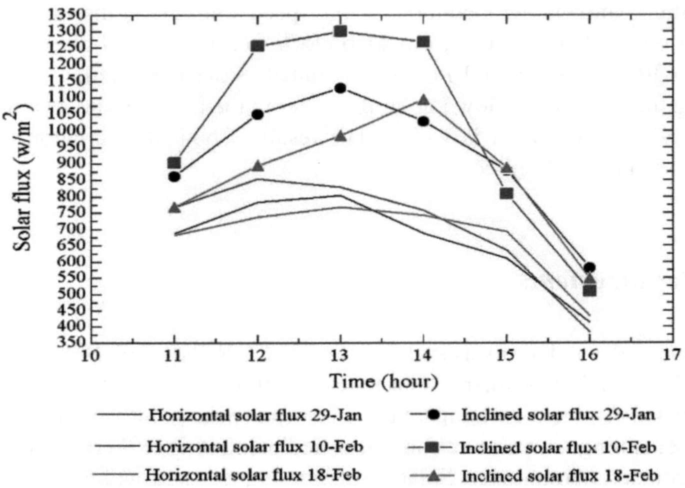

Fig. 3 Variation of insolation due to reflectors (Color figure online)

Fig. 4 Salinity effect on the product output

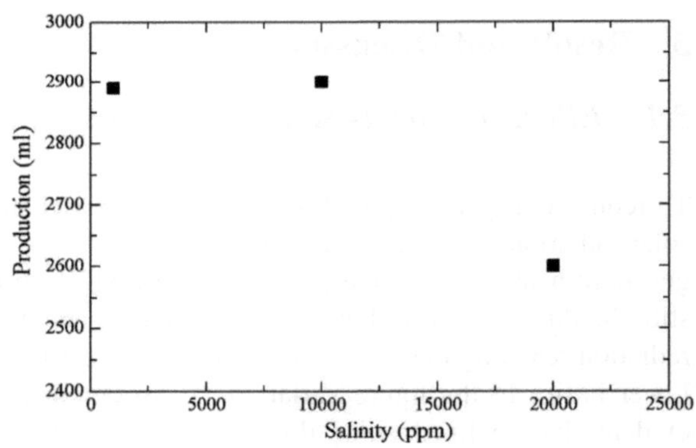

output decreases as in the steady water there is algae formation which reduces the rate of evaporation in the still resulting in lower product output (Fig. 4).

Another reason for lower product output is the salt accumulation in the basin which reduces the vapour pressure eventually reduces the product output. The total output of one day is shown for three different saline waters. It is evident that salinity does not affect the output of the system. Although in the 20,000 ppm saline water product output decreases, about 1%, but it is due to less intensity, about 10%, and less temperature difference between the basin water and glass.

3.3 *Effect of Flow Rate on Production*

Experiments are done by varying the flow rate of water in the upper basin. Flow rates of 10, 30, 50 and 70 ml/min are studied. Increasing the flow rate above 70 ml/min resulted in similar product output. The purpose of this study was to find the optimum flow rate with which we can find the maximum product output for the still. In the below figure the product output for a particular flow rate is shown. It is evident that the flow rate is increasing with increase in flow rate. The reason for that is as flow rate is increased, glass temperature of lower basin reduces which creates more temperature difference between lower basin and glass, and results in more product output of the lower basin (Fig. 5).

3.4 *Production Output with Time*

As discussed earlier the experiment was performed with a different salinity of the saline water; here, in the graph comparison of hourly output is shown. It is evident that pattern of the hourly output is similar although the salinity is different. In the 20,000 ppm case, the output is low due to low solar intensity. The production of the still was taken at every hour (Fig. 6).

3.5 *Upper and Lower Production with Time*

In the double effect still, both the lower basin and the upper basin yields distillate. It is seen that yield is more in the case of upper basin than lower basin due to direct sunlight. The lower basin is supplied with pre-heated water, and the upper basin is supplied the normal temperature water. Water in the upper basin reduces the temperature of the lower basin glass cover, and creates the temperature difference

Fig. 5 Flow rate effect on the product output

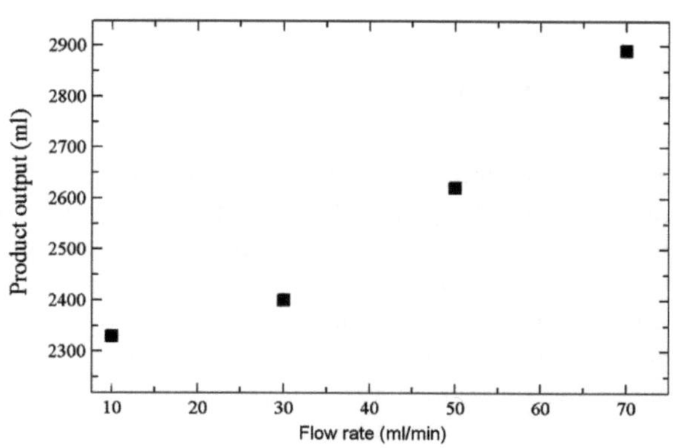

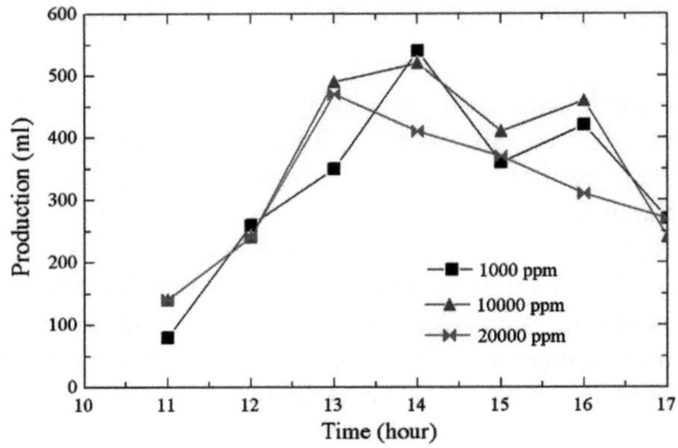

Fig. 6 Variation in product output with time for all data

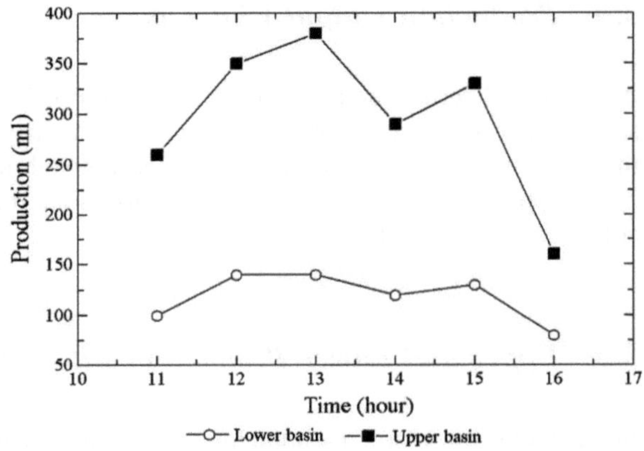

Fig. 7 Variation in product output of both the basin with time

between the basin water and glass which yields to distillate output. Here, in the graph a case study of 10,000 ppm water is taken. It is evident that output is more in the upper basin, but the output pattern is similar in both the basins (Fig. 7).

4 Conclusion

The present study represented a double effect solar still with north–south V-trough reflectors and flat plate collector. Otherwise, wasted latent heat of condensation is utilised to gain more product output. Despite salt concentration variation from 1000 to 20,000 ppm during the experiment, the product output remains constant. It shows that there is no salt deposition due to a continuous flow of salty water. It affects less maintenance requirement than the conventional solar still.

Acknowledgements CSIR-CSMCRI Communication No. PRIS/105. We are thankful to Dr. Pratap S. Bapat, Mr. R. Rathor, Mr. Jitendra Zala, and Mr. Jalpesh of Solar Energy Group, Process Design and Engineering Cell of CSMCRI for helpful assistant in work.

We are thankful to CSIR-CSMCRI for providing infrastructural facility.

References

1. S.A. El-Agouz, Y.A.F. El-Samadony, A.E. Kabeel, Performance evaluation of a continuous flow inclined solar still desalination system. Energy Convers. Manag. **101**, 606–615 (2015)
2. R. Kalbasi, M.N. Esfahani, Multi-effect passive desalination system, an experimental approach. World Appl. Sci. J. **10**(10), 1264–1271 (2010)
3. S.A. Kalogirou, Seawater desalination using renewable energy sources. Prog. Energy Combust. Sci. **31**, 242–281 (2005)
4. S. Maiti, C. Bhatt, P. Patel, P.K. Ghosh, Practical and sustainable household seawater desalination using an improved solar still. Desalin. Water Treat. 1–14 (2014)
5. T. Rajaseenivasan, K.K. Murugavel, Theoretical and experimental investigation on double basin double slope solar still. Desalination **319**, 25–32 (2013)

Transient Heat Transfer Analysis in Insulated Pipe with Constant and Time-Dependent Heat Flux for Solar Convective Furnace

Manish Sachdeva and Laltu Chandra

Abstract The paper deals with heat transfer during the transportation of the hot air from receiver via an insulated pipe to solar convective furnace for metal processing. In the developed concept of solar convective furnace (SCF) hot air at a temperature of 300–600 °C is required for a duration of about 3–6 h. However, the availability of solar irradiance is 8–10 h per day, whereas, the typical required time for achieving the steady condition is in the order of hours. Thus, prediction of time-dependent air temperature development at the outlet of insulated pipe is required for a given flow condition. Considering these aspects, the developed and validated transient heat transfer analysis tool is used to analyze effect of (a) pipe length and thickness of insulation, (b) mass-flow-rate of air, (c) constant and variable inlet air temperature and (d) preheating of pipe. Following are the broad observations based on the performed analysis: (i) time to reach steady state for the considered insulated pipe and the ratio of the heat loss to input reduces with increasing mass-flow-rate; (ii) increasing insulation thickness beyond critical thickness reduces the heat loss that results in a higher temperature at outlet. However, it increases the required time to reach the steady state condition; (iii) the achieved maximum temperature corresponding to a solar irradiance is forward shifted in time.

Keywords Insulated pipe · 1-D numerical tool · Experiment · Preheating · Steady state time · Insulation thickness

---

M. Sachdeva (✉)
Department of Mechanical Engineering, Indian Institute of Technology Jodhpur, Jodhpur 342011, Rajasthan, India
e-mail: sachdeva@iitj.ac.in

L. Chandra
Department of Mechanical Engineering and Center for Solar Energy, Indian Institute of Technology Jodhpur, Jodhpur 342011, Rajasthan, India
e-mail: chandra@iitj.ac.in

L. Chandra and A. Dixit (eds.), *Concentrated Solar Thermal Energy Technologies*, Springer Proceedings in Energy, https://doi.org/10.1007/978-981-10-4576-9_22

Nomenclature

A_a	Area of aperture (m^2)
C	Concentration factor
C_{pf}	Specific heat capacity of fluid (J/kg K)
C_{ps}	Specific heat capacity of pipe/absorber (J/kg K)
DNI	Direct normal irradiance (W/m^2)
E_a	Net energy available at inlet (MJ)
E_{ph}	Energy required for preheating the pipe (MJ)
E_u	Energy utilized (MJ)
h_f	Heat transfer coefficient for internal heat transfer in fluid (W/m^2 K)
h_{ex}	Natural convective heat transfer coefficient (W/m^2 K)
K_i	Thermal conductivity of insulation (W/mK)
K_s	Thermal conductivity of pipe (W/mK)
L_c	Thermal development length (m)
L_s or L	Length of pipe (solid domain) (m)
$\dot{m}_f$	Mass flow rate (kg/s)
n_{th}	Efficiency of receiver
$N_{uex,i}$	Nusselt number for natural convection at top of insulation domain (radial boundary)
$N_{uex,s}$	Nusselt number for natural convection at axial boundary of solid domain
N_z	Number of divisions in axial direction of pipe
P_{rf}	Prandtl number of fluid
$\dot{Q}_a$	Power available at the inlet of pipe (W)
$\dot{Q}_{cond}$	Power conducted (W)
$\dot{Q}_{fts}$	Power transferred from fluid to solid (W)
$\dot{Q}_{Nat.Conv}$	Natural convective power losses from insulation surface (W)
$\dot{Q}_{sti}$	Power transferred to insulation from solid domain (W)
r	Radial direction of pipe
Ra	Rayleigh number
Re	Reynolds number
r_i	Outer radius of insulation (m)
r_s	Inner radius of pipe (m)
SST	Time to reach steady state (s)
T	Time axis
T_a	Ambient temperature (K)
t_h	Time (h)
T_i	Temperature of insulation (K)
T_m	Mean temperature of fluid (K)
$T_{m,in}$	Mean temp of fluid at inlet of pipe (K)
$T_{m,out}$	Mean temp of fluid at outlet of pipe (K)
$T_{m,out\|steady\ state}$	Mean temp of fluid at outlet at steady state (K)
T_{ph}	Preheat temp of pipe (K)

T_s	Temperature of pipe (K)
t_{op}	Total time of operation of the system (s)
$T_{m_i}^n$	Mean temp of fluid at *i*th node and *n*th time step (K)
$T_{s_i}^n$	Temperature of solid domain at *i*th node and *n*th time step (K)
$T_{i_i}^n$	Temperature of insulation at *i*th node and *n*th time step (K)
z	Axial direction of pipe

Greek Symbols

α_i	Thermal diffusivity of fluid domain (m^2/s)
α_s	Thermal diffusivity of solid domain (m^2/s)
δ_i	Thickness of insulation (m)
δ_s	Thickness of pipe (m)
Δr	Grid spacing in radial direction (m)
Δz	Finite difference element in axial direction (m)
ø	Porosity of absorber
ρ_s	Density of pipe (kg/m^3)
Δz	Grid spacing in axial direction (m)
Δt	Time step (s)

Abbreviations

MFR	Mass flow rate
OVAR	Open volumetric air receiver
VNS	Von Neumann stability
Num	Numerical
POA	Power on aperture (W)
DNI	Direct normal irradiance (W/m^2)
Exp	Experiment

1 Introduction

Insulated pipe is used in transportation of working fluid in any solar thermal system for various applications. One such application is a retrofitted solar convective furnace (SCF) system [1, 2]. In this system, shown in Fig. 1a, an insulated pipe is used to transport hot air from an open volumetric air receiver (OVAR) [3–5] for metal processing. It is obvious that hot air will lose heat to the relatively cold pipe. A part of this energy heats up the pipe while the remaining is transferred to the insulation covering the pipe, which is typically made of Ceramic Wool (K_i = 0.15 W/mK). The heat is then conducted through the insulation and eventually lost to ambient via convection. This results in reduced air temperature at the pipe outlet, which is connected to the retrofitted furnace. As the pipe starts to heat

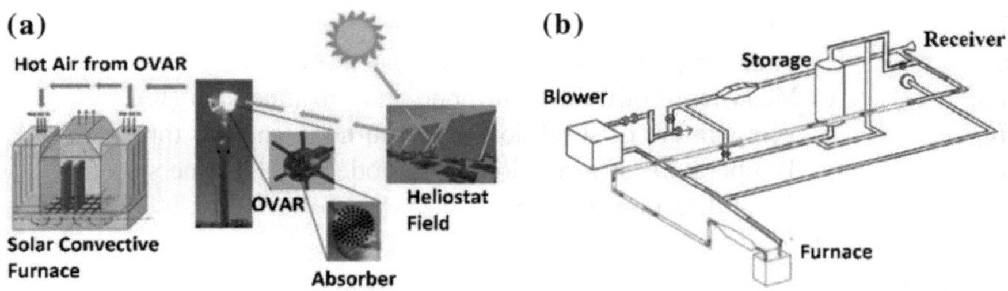

Fig. 1 **a** Working of solar convective furnace and **b** schematic of solar air tower simulator (SATS) facility

up, the outlet air temperature increases and reaches a steady state corresponding to a fixed inlet air temperature. The required time to achieve the desired steady state is of the order of hours [6, 7]. The required time depends on parameters, like mass flow rate, pipe length, insulation thickness, etc. It is to be noted that the availability of solar radiation is 8–10 h a day. Also, the direct normal irradiance (DNI), which is required for concentrated solar thermal systems, varies throughout the day leading to variable pipe outlet air temperature. Considering these aspects, an experimentally validated numerical tool [8] is used to analyze the effect of (a) pipe length, (b) insulation thickness, (c) preheating of pipe on development of air temperature at the pipe outlet for constant as well as time-dependent heat flux conditions.

2 Methodology

2.1 Mathematical Modeling: Insulated Pipe

The adopted numerical approach models the transport of hot fluid in an insulated pipe as, fluid (air), solid (pipe) and insulation (ceramic wool) domains. The details of which are as follows.

2.1.1 Fluid Domain

The control volume as in Fig. 2a depicts one-dimensional axial heat transfer in bulk fluid domain with heat loss to pipe in radial direction. This is modeled using conservation of energy as in Eq. (1). The storage term is neglected in comparison to the convective term owing to high mass flow rate. The heat transfer coefficient is estimated using a modified Dittus-Boelter correlation including variation of Nusselt number along the development length as in Eq. (2). The boundary condition is given constant or time-dependent inlet temperature. These are attributed to constant or time-dependent heat flux conditions.

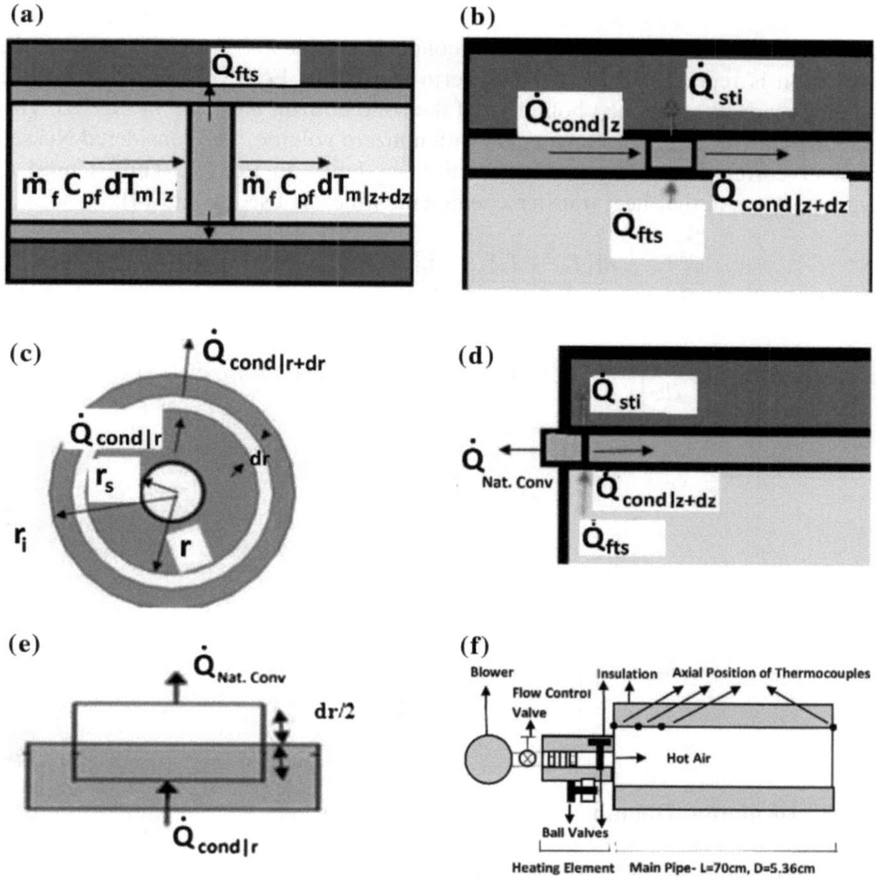

Fig. 2 Control volume (CV) showing heat transfer, **a** in fluid domain, **b** in solid domain, **c** in insulation domain, **d** at one end of the solid domain, **e** at boundary of insulation domain and **f** represents the schematic of experimental setup for validation

$$\dot{m}_{\mathrm{f}}\frac{\partial(c_{\mathrm{pf}}T_{\mathrm{m}})}{\partial z} = -2\pi r_{\mathrm{s}}h_{\mathrm{f}}(T_{\mathrm{m}} - T_{\mathrm{s}}) \tag{1}$$

$$Nu_{\mathrm{f}} = 0.027 * Re^{0.8}Pr_{\mathrm{f}}^{\frac{1}{3}}\left(\frac{z}{\mathrm{Lc}}\right)^{-0.0054}. \tag{2}$$

2.1.2 Solid Domain

The control volume in Fig. 2b depicts one-dimensional heat transfer in the bulk solid domain. Heat transfer in radial direction is neglected as Biot number

$(Bi = h_f \delta_s / k_s \sim 0.003)$ is substantially lower than 1. The heat transfer in solid domain is modeled with Eq. (3). The conductive heat loss from the solid to the insulation is represented by the third term in LHS of Eq. (3). The selected finite volume element at the inlet boundary of the solid domain is shown in Fig. 2d. This is modeled with Eq. (4) following [9] with nonzero volume. The considered Nusselt number correlation for natural convection is shown in Eq. (5). This is used to estimate the external heat transfer coefficient, which is used in Eq. (4).

$$\frac{\partial^2 T_s}{\partial z^2} + \frac{h_f (T_m - T_s)}{K_s \delta_s} + \frac{K_i}{K_s \delta_s} \frac{\partial T_i}{\partial r}_{(r = r_s + \delta_s)} = \frac{1}{\alpha_s} \frac{\partial T_s}{\partial t} \tag{3}$$

$$\left[(-1)^n 2\pi r_s \delta_s k_s \frac{\partial T_s}{\partial z}\right]_{z=0,L} + [\pi r_s dz h_f (T_m - T_s)]_{z=0,L} + \left[K_i \pi r_s dz \frac{\partial T_i}{\partial r}\right]_{r=r_s+\delta, z=0,L} - [h_{ex} 2\pi r_s \delta_s (T_s - T_a)]_{z=0,L} = \left[\pi r_s \delta_s dz \rho_s c_{ps} \frac{\partial T_s}{\partial t}\right]_{z=0,L} \tag{4}$$

where, $n = 1$ for $z = 0$, $n = 2$ for $z = L_s$.

$$Nu_{ex,s} = \left[0.825 + \frac{0.387 Ra^{1/6}}{\left[1 + (0.492/Pr_f)^{4/9}\right]^{8/27}}\right]^2 \tag{5}$$

2.1.3 Insulation Domain

The control volume for heat transfer in insulation domain is shown in Fig. 2c. The axial heat transfer is neglected in lieu of low thermal conductivity. The radial heat transfer is dictated by heat loss to ambient and conduction from solid insulation interface, which is represented by $T_s = T_i$ at $r = r_s$. The external boundary of insulation is depicted as a finite-element in Fig. 2e following [9], which is modeled using Eq. (7). Nusselt number correlation as in Eq. (8) [9] is used to estimate the used external convective heat transfer coefficient in Eq. (7).

$$\frac{\partial^2 T_i}{\partial r^2} + \frac{1}{r} \frac{\partial T_i}{\partial r} = \frac{1}{\alpha_i} \frac{\partial T_i}{\partial t} \tag{6}$$

$$\frac{1}{r_i} \frac{\partial T_i}{\partial r}_{(r=r_i)} + \frac{h_{ex} \left(T_{i(r=r_i)} - T_a\right)}{K_i} = -\frac{1}{2\alpha_i} \frac{\partial T_i}{\partial t}_{(r=r_i)} \tag{7}$$

$$Nu_{\mathrm{ex,i}} = \left[0.6 + \frac{0.387Ra^{1/6}}{\left[1+(0.559/Pr_{\mathrm{f}})^{4/9}\right]^{8/27}}\right]^2 \tag{8}$$

First-order explicit finite volume method is used to discretize the equations. Corresponding Von Neumann stability (VNS) conditions for the equations are satisfied. The fluid equation is unconditionally stable while Eqs. (9) and (10) represents the VNS equations for solid and insulation domains respectively. Further, experiments were performed for the validation of the numerical tool, the schematic of which is shown in Fig. 2b. For this purpose a constant mass flow rate and inlet temperature condition is employed. The experimental setup consists of two pipes, a small heating pipe and the main pipe. The pipes are connected through a T-joint and ball valve assembly. Operating the installed valves in the heating pipe a constant temperature of 400 K is achieved to the inlet of main pipe at a mass flow rate of 5 g/s. For further details on the modeling [8] may be referred.

$$\Delta t \le \frac{\Delta z^2/\alpha_{\mathrm{s}}}{2 + \frac{h_{\mathrm{f}}(\Delta z)^2}{K_{\mathrm{s}}\delta_{\mathrm{s}}} + \frac{K_{\mathrm{i}}(\Delta z)^2}{K_{\mathrm{s}}\delta_{\mathrm{s}}(\Delta r)} + \frac{2h_{\mathrm{ex}}(\Delta z)}{K_{\mathrm{s}}\delta_{\mathrm{s}}}} \tag{9}$$

$$\Delta t \le \frac{\Delta r^2/\alpha_{\mathrm{i}}}{2 + \frac{2\Delta r h_{\mathrm{ex}}}{K_{\mathrm{i}}}}. \tag{10}$$

2.2 Transport of Hot Air in an Insulated Pipe for SCF

As explained, hot air is to be transported through an insulated pipe for use in SCF. The requirement of SCF is based on applications, for example, annealing/soaking of different metals. Therefore, both constant and time-dependent heat flux conditions for the insulated pipe are of equal importance. In view of this, the developed and validated numerical tool is employed for analyzing these conditions. These are as follows.

2.2.1 Constant Heat Flux

For analyzing transport of hot air through the described insulated pipe in Fig. 1b, simulations are performed, as in experiment, with a constant inlet temperature. The considered parameters for this purpose are summarized in Table 1. Further, expression for the analyzed parameters, like, energy at the pipe inlet (E_{a}), energy required for preheating of pipe (E_{ph}) energy utilization by the reduced steady state time (E_{u}), heat gain rate $\left(\dot{Q}_{\mathrm{a}}\right)$ and the rate of heat loss to ambient $\left(\dot{Q}_{l}\right)$ are presented in Eqs. (11)–(15). For parametric analysis following are considered in view of the

Table 1 Parameters for numerical simulations with constant heat flux

L (m)	δ_i (mm)	T_{ph} (K)	$T_{m,in}$(K)	MFR (g/s)
8.75	23.6	300	673	10
4	23.6	300	673	10
8.75	47.2	300	673	10
8.75	23.6	375	673	10
8.75	23.6	450	673	10
8.75	23.6	300	673	5
8.75	23.6	300	673	15

installed SCF system at IIT Jodhpur, which are (a) $4 \leq \boldsymbol{L}\textbf{(m)} \leq 8.75$, (b) $23.6 \leq \boldsymbol{\delta_i}$ **(mm)** ≤ 47.2 with critical insulation thickness of ($K_i/h_{ex} \sim 23.6$ mm) [10], (c) $300 \leq \boldsymbol{T_{ph}}$ **(K)** ≤ 450 and (d) fixed inlet temperature $\boldsymbol{T_{m,in}}$ **(K) = 673** and (e) $5 \leq$ **MFR (g/s) of air** ≤ 15.

$$E_a = \dot{m}_f C_{pf} T_{m,in} t_{op} \tag{11}$$

$$E_{ph} = \rho_s(2\pi r_s \delta_s L_s) C_{ps}(T_{ph} - T_a) \tag{12}$$

$$E_u = \dot{m}_f C_{pf} T_{m,out}(t_{op} - \mathrm{SST}) - \rho_s(2\pi r_s \delta_s L_s) C_{ps}(T_{ph} - T_a) \tag{13}$$

$$\dot{Q}_a = \dot{m}_f C_{pf} T_{m,in} \tag{14}$$

$$\dot{Q}_l = \sum_{n=0}^{Nz} h_f 2\pi(r_s + \delta_s)\Delta z(T_m^n - T_s^n) \tag{15}$$

where, t_{op} is the operation time for the system = 8 × 3600 = 28,800 s. This parameter indicates the maximum duration of operation.

2.2.2 Time-Dependent Heat Flux

As described, the employed DNI for SCF varies throughout the day. Thus, the possible operating conditions are (A) constant MFR leading to a time-dependent temperature at the inlet of insulated pipe and (B) variable MFR leading to a constant temperature at the inlet of insulated pipe (Table 2). To investigate the heat transfer process under such a scenario, DNI data of Padova, Italy on 18 August from 9 am to 5 pm is considered, see Fig. 3a [11]. The selection is purely qualitative including the provided uncertainty. For simulation, a cubic fit of the considered DNI data is employed, which is given by Eq. (16) and is shown in Fig. 3a. The cubic fit has a root mean square error of 5.59 W/m^2 corresponding to maximum variation of $\sim$3% in DNI data. The required power-to-aperture (POA) is then estimated using Eq. (19) for area concentration factor (C) as 500. The POA is utilized to calculate either the time-dependent inlet temperature or MFR using an experimental correlation of

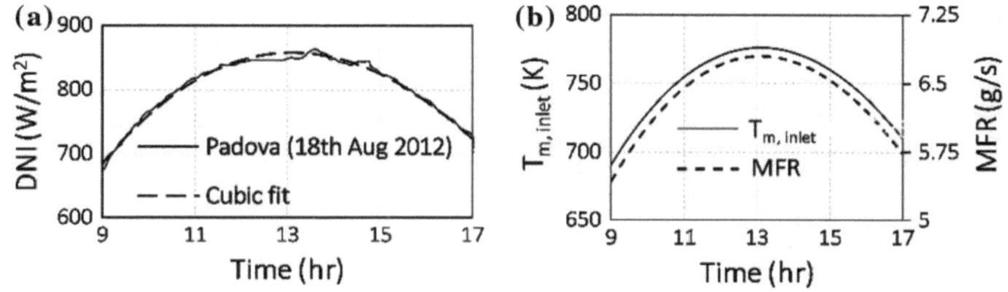

Fig. 3 **a** Direct normal irradiance data of Padova city as on 18th Aug 2016 [11] and the corresponding cubic fit curve. **b** Time-dependent inlet air temperature to pipe

Table 2 Details of the operating conditions

Operating condition →	Condition A	Condition B
MFR (g/s)	5	Variable
Inlet temperature (K)	Variable	673
L (m)	4	4
δ_i (mm)	23.6	23.6
T_s (t = 0) (K)	304	304

thermal efficiency of OVAR as in Eq. (17). The definition of thermal efficiency is given in Eq. (18).

$$\mathrm{DNI} = -1189 + 344.0\,t_h - 16.73\,t_h^2 + 0.1832\,t_h^3 \tag{16}$$

$$n_{th} = 0.994\left(\frac{\mathrm{POA}}{\dot{m}_f}\right)^{-0.083}(1 - \text{ø})^{-0.34} \tag{17}$$

$$n_{th} = \frac{\dot{m}_f C_{pf} \Delta T_m|_{\text{across receiver}}}{\mathrm{POA}} \tag{18}$$

$$\mathrm{POA} = \mathrm{DNI} * C * A_a. \tag{19}$$

3 Results and Discussions

3.1 Constant Heat Flux: Validation and Discussion

Figure 4 shows a comparison between the measured and numerically analyzed temperature developments in solid and fluid domains, respectively. Qualitatively the temperature development and the time to reach steady state are comparable. Quantitatively deviation of about 5–15% is observed. The overestimation of fluid

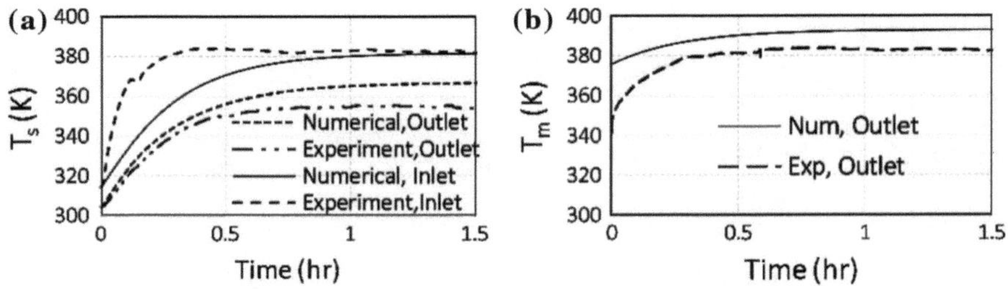

Fig. 4 Comparison between numerical and experimental temperature **a** in pipe at the inlet (z = 0) and outlet (z = 0.7 m), **b** in fluid at the outlet (z = 0.7 m)

Table 3 Time to reach steady state and energy utilized for different simulation conditions

L (m)	δ_i (mm)	T_{ph} (K)	$T_{m,in}$ (K)	$T_{m,out}$ (K)	MFR (g/s)	SST (s)	E_{ph} (MJ)	E_u (MJ)	t_{op} (s)	E_a (MJ)	E_u/E_a (%)
8.75	23.6	300	673	509.0	10	6364	0	117.8	28,800	207.0	56.9
4	23.6	300	673	583.4	10	4553	0	148.2	28,800	207.0	71.6
8.75	47.2	300	673	536.4	10	7164	0	120.4	28,800	207.0	58.2
8.75	23.6	375	673	509.0	10	6068	1.6	119.4	28,800	207.0	57.7
8.75	23.6	450	673	509.0	10	5647	3.2	121.6	28,800	207.0	58.8
8.75	23.6	300	673	434.8	5	8816	0	44.3	28,800	103.5	42.8
8.75	23.6	300	673	548.3	15	5026	0	204.8	28,800	310.5	66.0

temperature is attributed to underestimation of heat loss to ambient. Thus, forced or mixed convective based heat loss is expected to improve the prediction.

The effect of pipe length, insulation thickness, preheat temperature and air mass-flow-rate on the required steady state time (SST) and energy utilization (E_u) for 8 h of continuous operation is summarized in Table 3. Following can be inferred from these analyses:

1. As L increases, SST decreases and $T_{m,out}|_{steady\ state}$ increases result in a higher E_u.
2. With the increase in δ_i, both $T_{m,out}|_{steady\ state}$ and SST increases. The net result is increase in E_u.
3. As expected, preheating results in decrease of SST. Also, preheating implies energy input. However, the net energy utilization increases by around 4 MJ when the pipe is preheated to 450 K.
4. As MFR increases, SST decreases and E_u increases.

Figures 5 and 6 show the development of mean temperature of hot air at the outlet of pipe and the rate of heat loss as a function of time for the considered parameters. Figure 5 shows a stronger dependence of air outlet temperature and loss of power on the pipe length in comparison to insulation thickness. This is evident from the increase in outlet temperature of hot air for approximately similar changes

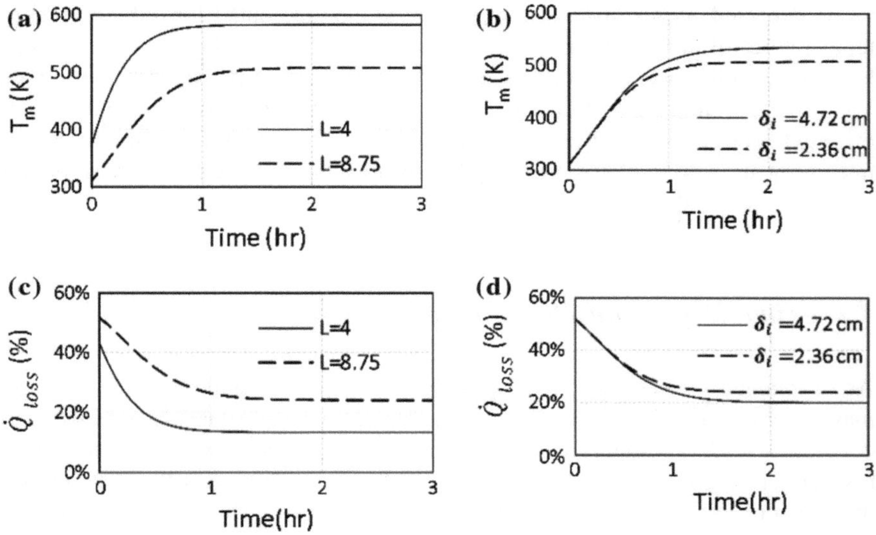

Fig. 5 **a**, **b** Temperature development at pipe outlet for different pipe length (m) and insulation thickness. **c**, **d** Percentage of net power loss for different pipe length and insulation thickness, respectively

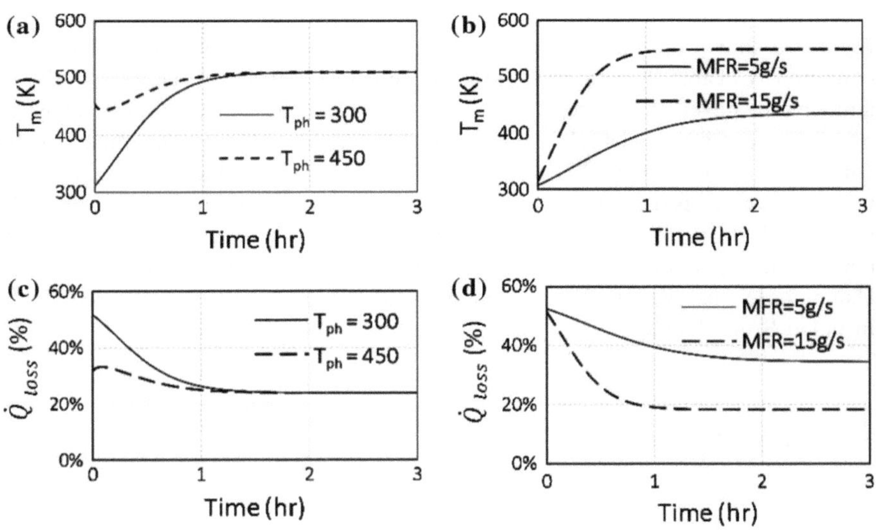

Fig. 6 **a**, **b** Air temperature development at the pipe outlet. **c**, **d** Power loss (in %) with time for different preheat temperatures and mass-flow-rate (MFR) respectively

in length and insulation. Thus, it can be inferred that reducing pipe length would provide higher energy utilization.

As expected, steady state air temperature at the pipe outlet for different preheat temperatures of pipe is observed to be the same as shown in Fig. 6a. The transient heat losses before reaching the steady state decrease with preheating; see Fig. 6c. Thus, the required SST decreases. This implies that the designed SCF can be operated over an extended duration. Figures 6b, d show that increasing mass flow rate of hot air in an insulated pipe length leads to a higher air temperature at the pipe outlet and reduced heat loss.

3.2 Time-Dependent Heat Flux: Discussion

As explained, there are two possible operating conditions (a) constant mass-flow-rate and (b) variable mass-flow-rate of hot air in an insulated pipe. These conditions lead to variable and fixed temperature at the inlet of insulated pipe, respectively. To investigate the effect of fixed and variable mass-flow-rate on heat transfer process, the ambient condition is employed. Thus, no preheating is employed. The considered thickness of insulation and pipe length is summarized in Table 2. These are analyzed in following subsections.

3.2.1 Condition A: Constant Mass-Flow-Rate

Figure 7a shows the air temperature development at outlet of insulated pipe for condition A (Table 2) with mass-flow-rate of 5 g/s as in [4, 5]. Obviously the air temperature never attains steady state for variable heat flux condition. Thus, an intermediate system, such as, thermal energy storage system is recommended for controlled operation of an SCF with steady inlet temperature. Furthermore, it can be inferred from analysis that the achieved maximum air temperature at the pipe outlet is forward shifted by about 40 min with reference to maximum of DNI. The temperature variation at the outlet between 11:00 and 16:30 h is less than 20 K. Using these observations the design of insulated pipe will be upgraded in the installed SATS facility.

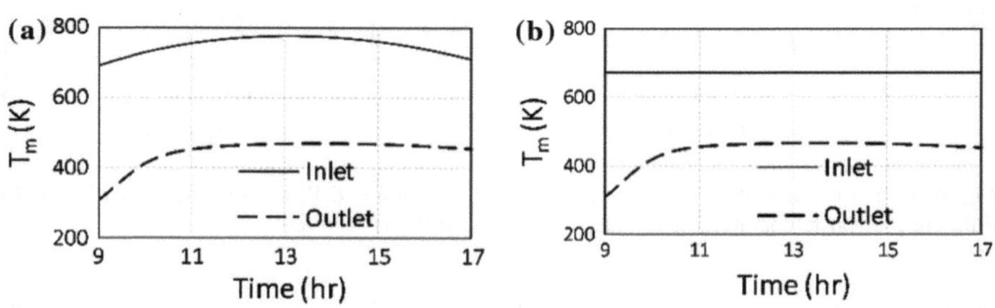

Fig. 7 Air inlet and outlet temperatures for **a** condition A and **b** condition B

3.2.2 Condition B: Time-Dependent Mass-Flow-Rate

Figure 7b shows the air temperature development at the pipe outlet for condition B (Table 2) with variable mass flow rate as depicted in Fig. 3b. Again it is observed that the temperature at the outlet of pipe does not reach the steady state, which is obvious. However, the maximum variation in temperature between 11:00 and 16:30 h is less than 11 K and a forward shift of about 24 min with respect to maximum DNI is observed. These are smaller than that of condition A. Thus, mode of operation as in condition A may be preferred over B. Even with this advantage, the need of thermal energy storage remains in order to operate the SCF in off hours.

4 Conclusion

The paper deals with transient heat transfer analysis during transport of hot air through an insulated pipe with constant and time-dependent heat flux condition. This is of practical relevance for operating a solar convective furnace. An experimentally validated numerical tool was used to study the effect of length of pipe, thickness of insulation and preheating at different mass flow rates for these conditions. As expected, the analysis showed that reducing the pipe length, increasing the insulation thickness beyond its critical limit and preheating of pipe helps in reducing heat losses. Thus enhancement of energy utilization is expected for a constant inlet air temperature. Further, the analysis with time-dependent inlet air temperature showed that the process never reached steady state. Thus, an intermediate system, thermal energy storage is recommended for operation of SCF under fixed and time-dependent mass flow rate modes of operation. Further, time-dependent mass flow rate operation is recommended due to low variation at the outlet temperature in some applications. Also, from the analysis performed, an insulated pipe system with an optimized length of ~4 m, insulation thickness ~60 mm is proposed for SATS facility.

Appendix

Equations (20), (21) and (22) represents the discretized form of the equations in fluid, solid and insulation domains and Eqs. (23) and (24) represents the discretized equations in solid and insulation domain respectively.

$$T_{\mathrm{m}_i}^{n+1} = T_{\mathrm{m}_i}^{n}(1 - ((b+1)/c) + T_{\mathrm{s}_i}^{n}(b/c) + T_{\mathrm{m}_i}^{n-1}/(c) \tag{20}$$

$$T_{\mathrm{s}_i}^{n+1} = T_{\mathrm{s}_i}^{n} * C(3) + T_{\mathrm{s}_{i+1}}^{n} * \tau_{\mathrm{s}} + T_{\mathrm{s}_{i-1}}^{n} * \tau_{\mathrm{s}} + T_{\mathrm{m}_i}^{n} * C(1) + T_{\mathrm{i}_{i,j=2}}^{n} * C(2) \tag{21}$$

$$T_{ii,j}^{n+1} = T_{ii,j}^{n} * g(2) + T_{ii,j-1}^{n} * \tau_i + T_{ii,j+1}^{n} * g(3) \quad (22)$$

$$T_{\mathrm{s}_i}^{n+1} = T_{\mathrm{s}_i}^{n} * D(4) + T_{\mathrm{s}_{i+1}}^{n} * \tau_{\mathrm{s}} + T_{\mathrm{s}_{i-1}}^{n} * \tau_{\mathrm{s}} + T_{\mathrm{m}_i}^{n} * D(1) + T_{\mathrm{i}_{i,j=2}}^{n} * D(2) + D(3) * T_{\mathrm{a}} \quad (23)$$

$$T_{\mathrm{i}_{i,\mathrm{Nz}}}^{n+1} = T_{\mathrm{i}_{i,\mathrm{Nz}}}^{n} * e(3) + T_{\mathrm{i}_{i,\mathrm{Nz}-1}}^{n} * 2\tau_{\mathrm{i}} + T_{\mathrm{a}} * e(2) \quad (24)$$

where

$$\begin{aligned} b &= \left(\frac{2\pi r_{\mathrm{s}} h_{\mathrm{f}} \Delta z}{m_{\mathrm{f}} c_{\mathrm{pf}}}\right), \mathrm{c} = \left(\frac{\rho_{\mathrm{f}} \pi r_{\mathrm{s}}^2 \Delta z}{m_{\mathrm{f}} \Delta t}\right), \alpha_{\mathrm{s}} = \frac{\rho_{\mathrm{s}} C_{\mathrm{ps}}}{K_{\mathrm{s}}}, \tau_{\mathrm{s}} = \frac{(\Delta z)^2}{\alpha_{\mathrm{s}} \Delta t}, C(1) \\ &= \frac{h_{\mathrm{f}} \tau_{\mathrm{s}} (\Delta z)^2}{K_{\mathrm{s}} \delta_{\mathrm{s}}}, C(2) = \frac{K_{\mathrm{i}} \tau_{\mathrm{s}} (\Delta z)^2}{K_{\mathrm{s}} \delta_{\mathrm{s}} (\Delta r)}, e(1) = \frac{2(\Delta r)^2 \tau_{\mathrm{i}}}{K_{\mathrm{i}} r_{\mathrm{i}}}, C(3) \\ &= 1 - 2\tau_{\mathrm{s}} - C(1) - C(2), D(1) = \frac{h_{\mathrm{f}} \tau_{\mathrm{s}} (\Delta z)^2}{K_{\mathrm{s}} \delta_{\mathrm{s}}}, D(2) \\ &= \frac{K_{\mathrm{i}} \tau_{\mathrm{s}} (\Delta z)^2}{K_{\mathrm{s}} \delta_{\mathrm{s}} (\Delta r)}, e(2) = h_{\mathrm{ex}} * e(1), D(3) = \frac{2\tau_{\mathrm{s}} h_{\mathrm{ex}} (\Delta z)}{K_{\mathrm{s}} \delta_{\mathrm{s}}}, D(4) \\ &= 1 - 2\tau_{\mathrm{s}} - D(1) - D(2) - D(3), \tau_{\mathrm{i}} = \frac{(\Delta r)^2}{\alpha_{\mathrm{i}} \Delta t}, e(3) \\ &= 1 - 2\tau_{\mathrm{i}} \frac{\Delta r}{r_{\mathrm{i}}} - \frac{2\Delta r \tau_{\mathrm{i}} h_{\mathrm{ex}}}{K_{\mathrm{i}}}, g(1) = \frac{\Delta r}{r}, g(2) \\ &= 1 - 2\tau_{\mathrm{i}} - g(1) * \tau_{\mathrm{i}}, g(3) = (1 + g(1)) * \tau_{\mathrm{i}} \end{aligned}$$

References

1. D. Patidar, S. Tiwari, P. Sharma, L. Chandra, R. Shekhar, Open volumetric air receiver based solar convective aluminum heat treatment furnace system. Energy Proc. **69**, 506–517 (2015)
2. D. Patidar, S. Tiwari, P. Sharma, R. Pardeshi, L. Chandra, R. Shekhar, Solar convective furnace for metals processing. JOM, 9 (2015). https://dx.doi.org/10.1007/s11837-015-1633-z
3. B. Hoffschmidt, F.M. Téllez, A. Valverde, J. Fernández, V. Fernández, Performance evaluation of the 200-kW$_{\mathrm{th}}$ HiTRec-II open volumetric air receiver. J. Sol. Energy Eng. **125**, 87–94 (2003)
4. P. Sharma, R. Sarma, L. Chandra, R. Shekhar, P.S. Ghoshdastidar, Solar tower based aluminum heat treatment system: Part I. Design and evaluation of an open volumetric air receiver. Sol. Energy **111**, 135–150 (2015)
5. P. Sharma, R. Sarma, L. Chandra, R. Shekhar, P.S. Ghoshdastidar, On the design and evaluation of open volumetric air receiver for process heat applications. Sol. Energy **121**, 41–55 (2015)
6. L.L. Vasiliev, Heat pipes in modern heat exchangers. Appl. Therm. Eng. **25**, 1–19 (2005)
7. J. Legierski, W. Bosuslaw, G. de May, Measurements and simulations of transient characteristics of heat pipes. J. Microelectron. Reliab. **46**, 109–115 (2006)

8. M. Sachdeva, D. Saini, L. Chandra, An experimentally validated numerical tool for unsteady heat transfer analysis in a porous absorber and in an insulated pipe of solar convective furnace system. Int. J. Therm. Sci. (under review-2016)
9. Y.A. Cengel, Heat and Mass Transfer, 4th edn. (McGraw Hill Company, New York, 2011)
10. F.P. Incropera, D.P. DeWitt, *Introduction to Heat Transfer*, 3rd edn. (Wiley, New York, 1996), pp. 93–99
11. S. Dugaria, A. Padovan, V. Sabatelli, D.D. Col, Assessment of estimation methods of DNI resource in solar concentrating systems. Sol. Energy **121**, 103–115 (2015)

Part VIII
Smart Grid and Policy Research

Impact of Different Solar Radiation Databases on Techno-economics of Concentrating Solar Power (CSP) Projects in Northwestern India

Ishan Purohit, Saurabh Motiwala and Amit Kumar

Abstract Bankable solar radiation data is one of the key technical barriers towards large-scale development and deployment of solar power in India. Depending on the technology (solar photovoltaic or solar thermal) the requirement of solar irradiance varies viz. the stationary solar PV works on global horizontal irradiance (GHI) and concentrating solar power (CSP) systems comprises direct normal irradiance (DNI). Till date the cumulative operational CSP projects are of around 225 MW capacities in the country out of more than 5400 MW operational grid connected solar power projects. Availability of long-term DNI data is a key challenge towards designing of a CSP project as there are no existing ground (measured) databases for long-term time series DNI in context of India. The project developers are therefore considering the satellite or interpolated databases for designing of the project. In the present study, the impact of DNI through different databases has been estimated on annual capacity utilization factor (CUF) and levelized cost of electricity (LCOE) of a parabolic trough collector (PTC) based CSP projects for two representative locations, i.e., Jodhpur (Rajasthan) and Bhuj (Gujarat) of northwestern part of India from its hot and dry climatic zone. The DNI values have been referred from ground measurements (IMD, NIWE), satellite source (NASA, NREL, and SWERA) and time series databases viz. Meteonorm (Mn) and SolarGIS for the selected locations. The conversion of all selected databases (static and dynamic) has been made in Typical Meteorological Year (TMY) in order to use in System Advisor Model (SAM) software. It has been estimated that the annual DNI variation for the selected locations has been observed from 3 to 30% (or more) with respect to estimated DNI from ground measurements of GHI and diffuse irradiance by Indian Meteorological Department (IMD). It has been estimated that the mutual variation of annual CUF in

I. Purohit
Lahmeyer International (India) Pvt. Ltd., Gurgaon 122002, India

S. Motiwala (✉)
Renewable Energy Engineering and Management, TERI University, New Delhi 110070, India
e-mail: saurabhmotiwala01@gmail.com

A. Kumar
TERI University, New Delhi 110070, India

L. Chandra and A. Dixit (eds.), *Concentrated Solar Thermal Energy Technologies*, Springer Proceedings in Energy, https://doi.org/10.1007/978-981-10-4576-9_23

within the range of 1.0–32.0%; however in the same line the mutual deviation in LCOE (based on the benchmark cost of CERC) has been observed from 1.0 to 44.0% over the selected representative locations. The study establishes the criticality of the selection of DNI database at the design stage of the project and hence for the comprehensive project evaluation.

Keywords Direct normal irradiance (DNI) · Concentrating solar power (CSP) · Levelized cost of electricity (LCOE)

1 Introduction

With an ambitious target of 175 GW of installed renewable capacity by year 2022; India has fixed its commitment towards promotion and use of renewable energy. In order to realize its goals, the target of Jawaharlal Nehru National Solar Mission (JNNSM) launched in 2010 has been increased five folds from 20 GW solar power capacity to 100 GW. The provision of consistent and affordable electricity is crucial for the economic and sustainable growth of a developing country like India. The per capita electric consumption of India stands at 744 which is one fourth of world average [1]. So, far fossil fuels contribute about 87% of total installed capacity in India [2].

The CSP system produces heat and further electricity using mirrors to concentrate DNI in order to heat a fluid to a temperature range of 400–1000 °C. The thermal energy of this fluid is used in thermodynamic cycle (mostly Rankine) so as to convert mechanical energy to electricity. The electricity thus produced is injected to the transmission grid. CSP plants with thermal energy storage (TES) or hybrid (i.e., integrated combined cycle) with natural gas are potential sources of dispatchable renewable power. The efficiency of typical CSP plant is around 16% (around 84% of total incident energy is lost during heat to electricity conversion), which is lower than that of conventional thermal power plants. The overall cost of systems are declining every year, however extensive research and development in TES would be the key factor in its deployment. At the end of 2014, the total global installed capacity was 4.4 GW. Spain, being the world leader in total installations of 2.3 GW, USA with 1.6 GW is second while India with 225 MW takes third position by the end of 2014 [3]. Around 300 MW installations are likely to be added in next 2 years by South Africa, Morocco, India, and Middle East taking the worldwide total greater than 5 GW [4]. In terms of construction and planning, markets in South Africa and Morocco grew rapidly in 2014 with 300 MW and 350 MW capacities under construction respectively. Out of the four technologies available commercially, PTC, central receiver system (CRS), linear Fresnel reflector (LFR) and Stirling dish. Only PTC technology seems to dominate the market with over 85% installations globally. As a part of research and demonstration Government of India has taken an initiative and implemented a demonstration CSP project of 1 MW capacity at National Institute of Solar Energy (NISE) comprising

PTC and LFR technologies for solar thermal field; but the project could not meet the expected techno-commercial targets and encourage the CSP market in the country. In the inception phase of National Solar Mission (NSM) in India ACME has implemented 2.5 MW CSP project in the state of Rajasthan based on CRS technology; but the performance of the plan has been observed too low mainly due to very less availability of DNI. Recently, India has commissioned Dhursar CSP plant and Megha Solar Plant of 125 MW (world largest on LFR technology) and 50 MW respectively which takes its total installed capacity to 225 MW. During the phase-I of NSM, seven CSP of 470 MW were allocated after signing PPAs (at an average tariff of Rs. 11.41/kWh) with NVVN and three projects of 30 MW through migration scheme. No allotment of CSP has been made after Phase-1 of NSM. After the experience of Phase-I of JNM, Ministry of New and Renewable Energy (MNRE) has shown interest to promote CSP segment via allowing gas for partial energy generation.

In context of India, unreliable potential estimation, limited or non-availability of bankable solar radiation database, large capital investments and financial risks and lack of indigenous manufacturers have hindered the effective installations of solar thermal projects [5]. The MNRE, Government of India (GoI) along with National Institute of Wind Energy (NIWE) has initiated Solar Radiation Resource Assessment (SRRA) project to provide investor grade ground data through around 113 stations across the country. NIWE offers short-term ground data of 113 stations which comprises significant uncertainty; however now the developers are relying of the high resolution satellite or interpolated data for CSP system designing. The present study aims to explore the impact of mutual variation of available DNI databases in India on the techno-commercial performance of CSP projects. The exercise has been focused on two locations namely Jodhpur (Rajasthan) and Bhuj (Gujarat) which are considered most potential zones for CSP projects.

2 The Proposed Work

In this study, an attempt has been made towards estimation of variation in the techno-commercial feasibility of CSP project using all available DNI databases in context of India. Two representative locations namely Jodhpur (Rajasthan) and Bhuj (Gujarat) have been selected which falls on the Northwestern parts of the country comprising maximum potential of CSP based power generation. In context of India the long term measured DNI data is negligibly available for any potential location; however it could be derived through measured global and diffuse irradiance. Satellite data of DNI is available through several databases of various resolutions in static and time series formats. The qualitative inter-comparability of the databases has been presented in Table 1 [6–8].

Table 1 Features of different solar radiation databases in context of India

Database of	Source	Type of database	Remarks
Indian Meteorological Department (IMD), GoI	Ground	Static (monthly average daily)	Handbook of solar radiation data for India, 1981-solar radiation data for 18 stations in the country with measurement period ranging from 2 to 21 years [9]. Solar radiation over India, 1982, Derived radiation data for 145 stations
SRRA, NIWE	Ground		Solar radiation data along with weather parameters available for 115 stations across the country
National Aeronautics and Space Administration (NASA)	Satellite		Global solar radiation data (1° × 1°) available for 22 years; from 1985 to 2005 [10]
Solar and Wind Energy Resource Assessment (SWERA)	Satellite		Solar radiation data available based on climate data from 1985 to 1991 available with a spatial resolution of 40 km × 40 km [11]
National Renewable Energy Laboratory (NREL)	Satellite	Dynamic—time series TMY data	Hourly annual solar radiation data available with other meteorological parameters in CSV format with a resolution of 10 km × 10 km [12]
Meteonorm 6.0, 7.0, 7.1	Interpolated		Approximately 20 years solar radiation data available in Typical Meteorological Year (TMY) format and can be easily imported for simulations in softwares like TRNSYS, SAM, PVsyst, Energy Plus etc. [13]
SolarGIS	Satellite		High resolution climatic database which provides 15 min instantaneous values as well 60 min average values

3 Methodology

All the solar radiation databases (presented in Table 1) have been collected for both the representative locations; i.e., nine databases for Jodhpur and six databases for Bhuj and further imported in PVsyst software [14] for converting static to dynamic data. Solar radiation databases could be categorized in two key categories namely static and dynamic. Static databases are those which contain long-term monthly average daily values to represent solar radiation pattern over a selected location. These databases present solar radiation as 12 monthly values only based on long-term statistical analysis. The published data of IMD, NASA, SWERA, and SolarGIS are under this category for India. The limitation of such databases include

Fig. 1 Methodology flow chart

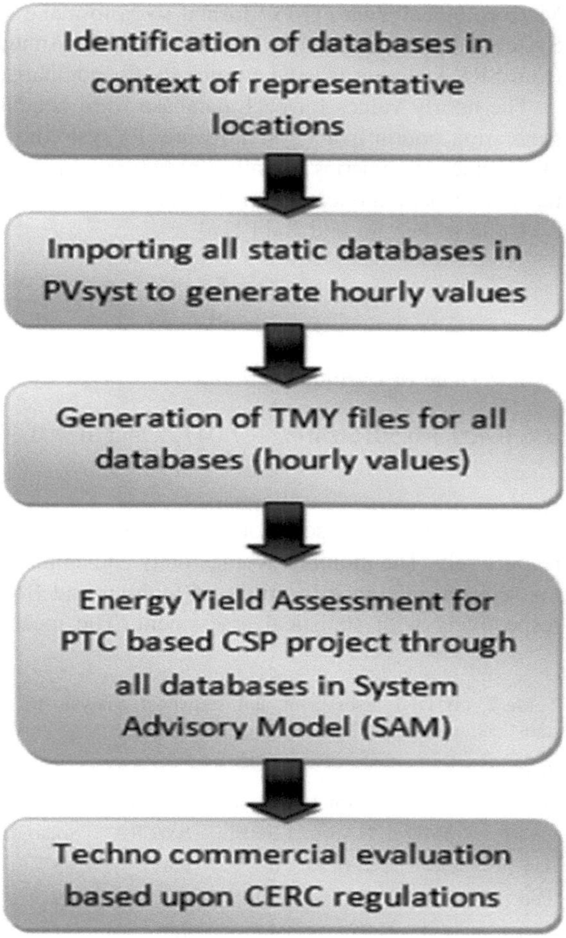

incapability to measure the instantaneous performance of the system, scheduling of the expected energy generation from the project, uncertainty associated and inter annual variability. A flow chart indicating step by step approach is being presented in Fig. 1.

The dynamic databases are those which presents hour-to-hour pattern (time series) of solar irradiance and associated meteorological based on long-term available data for a reference location. The databases address hour-to-hour energy generation pattern, scheduling, time series (transient) system analysis, uncertainty and inter annual variability, etc. The databases Meteonorm, NREL, SolarGIS, 3TIER, ISHRAE, etc., are under this category. In order to make the symmetry on system analysis all databases have been converted into hourly values using PVsyst software. The hourly values thus obtained were converted to standard Typical

Meteorological Year (TMY) format so as to make it compatible for simulations in System Advisor Model (SAM) software [15]. Annual average DNI value obtained from SRRA station was taken directly for comparison during analysis.

The hourly values for each database were obtained using synthetic hourly data generation option in PVsyst software. PVsyst converts imported monthly average DNI value into a series of daily irradiation values and then 24 values for each day using stochastic models and mathematical equation presented in Eq. (1) developed by Collares-Pereira and Rabl [16].

$$r_h(\omega, \omega_s) = \frac{\pi}{24}(a + b\cos\omega)\frac{\cos\omega - \cos\omega_s}{\sin\omega - \frac{\pi\omega_s}{180}\cos\omega_s} \tag{1}$$

The value of coefficients a and b are given by

$$\mathrm{a} = 0.409 + 0.5016(\sin\omega_s - 1.047) \quad \text{and} \quad \mathrm{b} = 0.6609 - 0.4767(\sin\omega_s - 1.047), \tag{2}$$

where ω and ω_s represent hour angle (in radian) and sunset hour angle (in radian) respectively. The monthly average daily and annual pattern of all DNI databases for the selected reference locations of Jodhpur and Bhuj are presented in Table 2a, b respectively with statistical assessment. The mean percentage value (MPE) and

Table 2 (a) DNI assessment and statistical analysis for Jodhpur, (b) DNI assessment and statistical analysis for Bhuj

(*a*)								
Month	DNI (kWh/m^2) databases over Jodhpur through							
	IMD	NASA	NREL	SWERA	SolarGIS	Mn 6.0	Mn 7.0	Mn 7.1
Jan	7.7	5.3	6.7	6.4	5.3	7.1	5.5	5.8
Feb	8.1	5.2	7.1	6.9	5.9	7.2	5.7	6.4
Mar	6.8	5.3	7.2	7.1	6.1	7.2	5.2	6.4
Apr	7.1	5.4	6.7	6.6	5.8	6.6	3.9	6.1
May	6.9	5.3	6.8	6.9	5.1	6.4	4.0	6.0
Jun	5.4	4.4	5.0	5.1	4.4	5.2	3.8	4.4
Jul	3.8	3.1	3.6	3.5	2.5	1.8	2.3	2.8
Aug	4.3	3.3	4.1	3.7	2.9	2.1	2.4	3.0
Sept	6.1	4.3	5.9	5.6	4.9	4.2	4.6	4.6
Oct	8.5	4.9	7.3	7.2	6.2	7.1	5.8	6.2
Nov	7.9	5.2	7.0	6.7	5.0	6.7	5.7	5.1
Dec	7.2	4.5	6.3	6.2	4.9	6.5	5.7	5.7
Avg.	6.6	4.7	6.1	6.0	4.9	5.7	4.5	5.2

(continued)

Table 2 (continued)

(*a*)								
Month	DNI (kWh/m^2) databases over Jodhpur through							
	IMD	NASA	NREL	SWERA	SolarGIS	Mn 6.0	Mn 7.0	Mn 7.1
Annual	**2427**	**1711**	**2240**	**2187**	**1801**	**2070**	**1658**	**1898**
MPE[a]	–	**−29.5**	**−7.7**	**−9.9**	**−25.8**	**−15**	**−31.8**	**−22**
(R^2)[b]	–	**0.75**	**0.91**	**0.87**	**0.8**	**0.87**	**0.84**	**0.8**
ASHs	**3849**	**3291**	**3434**	**3755**	**3396**	**3428**	**3530**	**3469**

(*b*)						
Month	DNI (kWh/m^2) databases over Bhuj through					
	IMD	NASA	NREL	SWERA	SolarGIS	Mn 7.0
Jan	6.7	5.1	5.5	7.0	5.8	7.3
Feb	6.8	4.9	6.6	7.3	5.8	6.3
Mar	5.9	4.9	6.6	7.5	6.2	7.2
Apr	6.5	4.9	6.2	7.5	6.7	6.2
May	7.0	4.8	6.1	6.9	5.9	5.8
Jun	4.8	4.0	5.4	4.8	3.9	4.6
Jul	2.4	2.6	2.1	2.6	1.8	2.2
Aug	2.6	2.5	1.9	2.7	1.9	2.2
Sept	4.7	3.6	4.4	5.2	3.9	4.5
Oct	6.7	4.5	6.5	7.1	5.7	6.3
Nov	7.0	4.8	5.5	7.0	5.2	6.6
Dec	6.8	5.2	5.7	6.8	5.3	6.5
Avg.	5.7	4.3	5.2	6.0	4.9	5.5
Annual	**2066**	**1575**	**1904**	**2206**	**1772**	**2004**
MPE	–	**−23.7**	**−7.8**	**6.8**	**−14.2**	**−3.0**
(R^2)	–	**0.93**	**0.84**	**0.92**	**0.87**	**0.88**
ASHs	**3572**	**3295**	**3469**	**3672**	**3285**	**3566**

[a]The MPE (mean percentage error) has been calculated using the expression; $\frac{\left[\sum\left\{\frac{I_{obs}-I_{pred}}{I_{obs}}\times 100\right\}\right]}{n}$

[b]R^2; coefficient of determination value determines goodness of fit of a particular model. Higher the R^2 value, higher would be its reliability

correlation coefficient (R^2) have also been presented with the estimation of annual sunshine hours (i.e., ASHs when DNI > 120 W/m^2) over the locations.

4 Technical Parameters for Energy Yield Assessment in SAM

In order to analyze the impact of DNI on technical performance of CSP projects a simulation study has been carried out for most commercial CSP technology, i.e., PTC for both selected locations using SAM software towards estimation of annual energy generation and capacity utilization factor (CUF). From the solar resource assessment exercise all databases have been converted in TMY through PVsyst and imported in SAM environment. The key design specifications of simulated PTC based CSP projects have been presented in Table 3. The considerations towards solar multiple, heat transfer fluid, collector and receiver tubes have been made based on best industrial practices.

Table 3 Technical specifications of PTC based CSP project

Solar field specifications		Collector specifications	
Plant capacity (MW)	1.0	Manufacturer	EuroTrough ET150
Degradation\year (%)	0.25	Reflective aperture area (m^2)	470.3
System availability (%)	96	Length of collector assembly (m)	100
Solar multiple	1.3	Receiver model	Schott PTR70 2008
No of loops	2	Absorber tube inner diameter (m)	0.066
Aperture area (m^2)	7524.8	Absorber tube outer diameter (m)	0.07
Cycle conversion (%)	37.74	Design gross output (MWe)	1.1
Thermal storage (h)	0	Gross to net conversion factor	0.9
Heat transfer fluid	Therminol VP − 1		
Inlet temp (°C)	293		
Outlet temp (°C)	391		
Pressure (bar)	100		

Table 4 Financial parameters of solar thermal projects as suggested by CERC

Parameter	Description
Capital cost (per MW)	₹120 million
Useful life	25 years
Debt:equity ratio	70:30
Interest rate on loan	12.70%
Return on equity (ROE)	20% for first 10 years
	24% from 11th year
Discount rate	10.67%
Depreciation	5.83% for first 12 years
	1.54% from 13th year
Working capital	1 month O&M, 15% of O&M expenses, 2 months of receivables
Interest on working capital	13.2%
O&M expenses (per MW)	₹1.677 million
O&M expenses escalation	5.72% per year

5 Financial Assessment Parameters

From the point of analyzing the effect of DNI variance in different databases on financial viability of CSP projects the financial assessment has been carried out via developing a techno-commercial model for estimation of LCOE.[1] The financial assumptions have been made in line with the benchmarking made by Central Electricity Regulatory Commission (CERC) for the financial year 2016–17 [17] as presented in Table 4.

6 Results and Discussion

It has been well established from the inter-comparability of monthly average and annual DNI that at a selected location there is significant variation in between the available DNI databases which strongly impacts the techno-commercial viability of the CSP project. Table 5 presents the overall techno-commercial findings of the

[1]The value of LCOE from CSP project can be determined by using the mathematical expression:

$$LCOE_{CSP} = \left[\frac{Ccsp\ CRF + Com,csp}{(365 \times 24) CFcsp\ Pcsp}\right]$$

where Ccsp represents capital cost of the CSP plant, CRF the capital recovery factor, Com, csp the annual operation, repair and maintenance cost of the CSP plant, *P*csp the capacity of the CSP plant and the capacity factor CRF is expressed as $CRF = \left\{\frac{d(1+d)^t}{d(1+d)^t - 1}\right\}$;
where d represents the discount rate and t represents the useful lifetime of the CSP project.

Table 5 Comparison of annual generation, CUF and LCOE for different databases

	Parameters	IMD	NASA	NREL	SWERA	SolarGIS	Mn 6.0	Mn 7.0	Mn 7.1
Jodhpur	Annual energy generation (MWh)	2334	1614	2528	2189	1766	2193	2033	1843
	CUF (%)	26.9	18.6	29.2	25.2	20.4	25.3	23.4	21.3
	MPE in CUF	–	−30.86	8.55	−6.32	−24.16	−5.95	−13.01	−20.82
	LCOE (₹/kWh)	9.37	13.55	8.65	9.98	12.37	9.97	10.75	11.85
	MPE in LCOE (%)	–	44.61	−7.68	6.51	32.02	6.40	14.73	26.47
Bhuj	Annual energy generation (MWh)	2,072	1,537	2,090	2,193	1,799	NA	2,164	NA
	CUF (%)	23.9	17.7	24.1	25.3	20.8	NA	25	NA
	MPE in CUF	–	−25.94	0.84	5.86	−12.97	NA	4.6	NA
	LCOE (₹/kWh)	10.55	14.22	10.46	9.96	12.14	NA	10.10	NA
	MPE in LCOE (%)	–	34.79	−0.85	−5.59	15.07	NA	−4.27	NA

study. From resource assessment it has been observed that the estimated annual DNI through IMD (1980) source is higher than 2400 kWh/m^2 which is comparable to best suited locations for CSP projects viz. California (USA), Spain, etc. The MPE w.r.t. IMD data of other databases varies from 3.0 to 31.8%. Simultaneously the annual CUF of the project varies from 18.6 to 29.2% for Jodhpur and 17.7 to 25.3% for Bhuj. The mutual variance has been estimated as 5.9 to 30.9 for Jodhpur and 0.8 to 25.9 for Bhuj. Additionally, LCOE varies from 8.65 to 13.55₹/kWh for Jodhpur and 9.96 to 14.22₹/kWh for Bhuj with a MPE of 6.4 to 44.6% and 0.9 to 34.8 respectively as compared to IMD. Even if the estimated DNI from IMD data is not considered a base case the mutual variation over the techno-commercial performance of CSP projects has been observed considerable. The time series databases shows less variation as compared with the satellite databases of low resolution.

7 Way Forward

DNI is more intermittent and site specific as compared with GHI; hence CSP projects could not be implemented anywhere like solar PV. In line with wind sector; MNRE needs to set up an institution (like NIWE for wind) for certification or ranking of solar radiation data. The short-term DNI of NIWE must comprise the uncertainty level for developers could estimate the energy generation at different probabilities viz. P75 or P90 for bankability.

References

1. The World Bank, Electric power consumption. See: http://data.worldbank.org/indicator/EG.USE.ELEC.KH.PC/countries/1W-IN?display=graph (2012)
2. CEA, Central Electricity Authority, All India installed capacity (in MW) of power stations. See: http://cea.nic.in/reports/monthly/installedcapacity/2016/installed_capacity-01.pdf (2016)
3. REN21, Renewables global status report. Renewable energy policy network for the 21st century (REN21), Paris. See: http://www.ren21.net (2015)
4. Greenpeace International, *European Solar Thermal Electricity Association (ESTELA), SolarPACES, Solar Thermal Electricity: Global Outlook 2016* (Greenpeace International, Amsterdam, 2016)
5. CSE, State of Concentrated Solar Power in India: A Roadmap to Developing Solar Thermal Technologies in India. Centre for Science and Environment, New Delhi (2015)
6. I. Purohit, P. Purohit, Techno-economic evaluation of concentrated solar power generation in India. Energy Policy **38**, 3015–3029 (2010)
7. I. Purohit, P. Purohit, S. Shekhar, Evaluating the potential of concentrating solar power generation in Northwestern India. Energy Policy **62**, 157–175 (2013)
8. I. Purohit, P. Purohit, Inter-comparability of solar radiation databases in Indian context. Renew. Sustain. Energy Rev. **50**, 735–747 (2015)
9. A. Mani, S. Rangarajan, *Solar radiation over India* (Allied Publishers, New Delhi, 1981)
10. NASA, Atmospheric Science Data Center. https://eosweb.larc.nasa.gov
11. SWERA, Solar and wind energy resource assessment. A global database developed by United Nation Environment Program. See: https://maps.nrel.gov/swera/ (2004)
12. NREL, India Solar Resource Data: Hourly Data and TMYs. See: (http://rredc.nrel.gov/solar/new_data/India/nearestcell.cgi. Accessed Feb 01 2016
13. Meteotest, Meteonorm database, See: www.meteonorm.com (2016)
14. PVsyst version 6.42. Photovoltaic Software. http://www.pvsyst.com/en/ (2016)
15. SAM 2016.3.14, System advisory model. National Renewable Energy Laboratory (NREL), See: https://sam.nrel.gov/download (2016)
16. M. Collares-Pereira, A. Rabl, The average distribution of solar radiation-correlations between diffuse and hemispherical and between daily and hourly insolation values. Sol. Energy **22**(1), 155–164 (1979)
17. CERC, Determination of benchmark capital cost norm for solar PV power projects and solar thermal power projects applicable during FY 2016–17. Central Electricity Regulatory Commission (CERC), Government of India, New Delhi. See: http://www.cercind.gov.in/2015/orders/SO17N.pdf

Envisioning Post Domain of Restructured Accelerated Power Development and Reforms Programme-Fault Current Limiters

Jagdish Prasad Sharma and Vibhor Chauhan

Abstract In the restructured environment, distributed generation (DG) can play an important role to maintain the expected voltage level, an improvement in network flexibility, sustainability, and reliability. The DG integration generally raises the fault level in many places, which may often exceed from withstanding capacity of present infrastructure. This increased fault level has the severe impact on power quality and voltage stability. Therefore, it is needed to operate existing network with the increased fault level constraint. With the Indian perspective of growing economy and increasing electricity demand, the fault current limiter (FCL) is a promising option to enhance the useful life of the equipments, even though fault current withstands capability has exceeded. These FCL have a modular design and easily customized as per constraints, system parameter, location choice, and space availability. A superconducting fault current limiter (SFCL) and solid-state fault current limiter (SSFCL) are two types of commercially available FCL solutions. At present, advancement in the high-temperature superconductor has reduced the cooling cost of SFCL, whereas high power switching development makes SSFCL more viable. There is a significant progress towards SFCL technologies across the global. To study various challenges and components of technology development for SFCL, a joint research project initiated by Crompton Greaves R&D and Ministry of Power, Government of India under the National Perspective Plan (NPP) project scheme. The focus of this paper is at describing how SSFCL may be considered for voltage sag mitigation and suppression of fault current for unsymmetrical/symmetrical faults on a parallel distribution feeder.

Keywords Voltage sag mitigation · Superconducting fault current limiter · Solid-state fault current limiter · Prospective fault current

J.P. Sharma (✉)
JK Lakshmipat University, Jaipur, India
e-mail: jpsharma.cseb@gmail.com

V. Chauhan
Jaipur Institute of Technology, Jaipur, India
e-mail: vibhor.chauhan@gmail.com

L. Chandra and A. Dixit (eds.), *Concentrated Solar Thermal Energy Technologies*, Springer Proceedings in Energy, https://doi.org/10.1007/978-981-10-4576-9_24

1 Introduction

In the present scenario, India power sector is going through renovations, modernization, and restructuring process to bridge the gap between demand and supply with enhancement of power quality, reliability, and customer satisfaction aspects. According to Central Electrical Authority ([1], March), there is an energy shortage of about 1.87% and a peak deficit of about 1.92% as on 30 January 2016. Bridging this gap from the generation side is constrained due to limited source and increased cost of new generation capacity addition. Apart from above, about 21–23% energy is lost in the transmission and distribution of power, which is very significant. The major part of this Technical and Distribution (T&D) loss associated with distribution sector, which is also an important link in the entire power sector. Distribution sector is an daunting sector due to its complexity, large public interface with different needs and expectations. These high losses are neutralizing all improvement initiatives. The reported technical and distribution (T&D) loss and aggregate technical and commercial (AT&C) loss are shown in Table 1 [1].

As distribution sector is constituted the cash register for the entire power system and government of India has initiated following sponsored schemes for distribution sector performance improvement as demonstrated [2].

- Integrated Power Development Scheme (IPDS)
- Deendayal Upadhyaya Gram Jyoti Yojana (DDUGJY)
- National Electricity Fund (NEF)
- Financial Restructuring Scheme

Both public and private power electrical utilities are coming forward to meet power shortage/power quality issues using distributed generation (DG) sources and energy-efficient technologies. DG interconnection improves the voltage profile and increases the fault current level, which is directly proportional to their penetration level. This makes existing protection system more vulnerable. Indian electrical utilities are restricted not too upscale or replacing the existing protection systems due to poor financial viability.

In this context, a fault current limiter (FCL) is a modern approach to increase existing asset utilization and fewer reinforcement requirements with enhancement of network flexibility, sustainability, and reliability. Modular design, voltage sag mitigation, and fast post-fault recovery voltage make solid-state fault current limiter (SSFCL) technology more suitable for DG integration below than 132 kV operating voltages. A superconducting fault current limiter (SFCL) is well suited for DG integration on or above than 132 kV operating voltages. This paper discusses the

Table 1 T&D and AT&C losses (%)

Component	2010–11	2011–12	2012–13	2013–14
T&D losses	23.97	23.65	23.04	21.46 (provisional)
AT&C losses	26.35	26.63	25.38	22.70 (provisional)

application of solid-state fault current limiter (SSFCL), which is useful to minimize increased fault level with existing protection system for the DG integrated distribution system.

2 Fault Current Limiter Technology

Looking to the Indian power sector, the fault current limiter (FCL) is an altered way to enhance the working life of equipment even if the fault level exceeds from their withstanding capability. FCL is also adaptable in terms of modularity, location choice, and space availability. The key benefits of fault current limiter are control of steady-state and prospective fault current with high reliability. A negligible impedance provides by FCL under normal condition and offers high impedance under faulty conditions. Generally, fault current limiting device location is in line with a transformer, in between neutral and ground of transformer connection and in a bus coupler. They are of two types: superconducting fault current limiter (SFCL) and solid-state fault current limiter (SSFCL). Under fault conditions, SFCL introduces high impedance by the extremely rapid loss of superconductivity, whereas SSFCL uses high-speed solid-state switching devices to rapid insertion of impedance to limit the fault current. For medium voltage network, it is also observed that the SSFCL have good in terms of fault current limiting capacity, restoration time and voltage sag mitigation parameter. Yadav et al. [3], have presented a qualitative comparison for FCL in terms of response and cost.

For technology development of a superconducting fault current limiter (SFCL), Crompton Greaves Ecumenical R&D and Ministry of Power, Govt. of India is conducting a joint research project under the National Perspective Plan (NPP). In the first phase of this project a 400 V, 800 A, 3 Phase resistive type SFCL had developed and the second phase of this project a 11 kV, 1250 A, 3 Phase resistive type SFCL are under process. The total cost of this project is approximately 306.00 (Rs. lakhs). The obtained response time, limited current, prospective, current and recovery time useful for the development of commercial SFCL for power grids [4].

As National Perspective Plan (NPP) is aimed to identify bridge technology gaps in various subsections of the Indian Power Sector through research and development. With consideration of overall cost, a majority of distributed generations is connected to 11 and 33 kV networks. This raised fault level at that point of coupling (PCC) in the network. Qualitative comparison of commercially reveals that the solid-state fault current limiter is more suitable to below 132 kV operated networks. To rejuvenate the distribution sector, most of the state utilities prioritized IPDS scheme for information technology adaptation and strengthening of the distribution sector. IPDS forward the targets laid down under R-APDRP for 12th & 13th plans and envisage the implementation of ERP & information technology adaptation in non-RAPDRP towns. The introduction of fault current limiters is an envisioned post domain of Restructured Accelerated Power Development and Reforms Programme (R-APDRP). For technology development of solid-state fault current limiter (SSFCL) is required to minimize increased fault level in the distribution system for adaptation to Indian conditions.

3 Role of Solid-State Fault Current Limiter

In this paper solid-state fault current limiter (SSFCL) utilized for voltage sag mitigation and fault current control. The basic topologies of solid-state fault current limiters are as under.

- Series switch type,
- Bridge type and
- Resonant type.

In this paper, we have utilized series switch type with inductor, which is inserted in series with a circuit. This series switch type SSFCL are composed of three parallel circuits as shown in Fig. 1. In case of fault the current path is diverted from Thyristor branch 1 to Thyristor branch 2 and overvoltage absorption implemented with the help of surge arrester and associated snubber circuit with Thyristor branches.

Control and current limiter Simulink model for above solid-state fault current limiter (SSFCL) depicted in Figs. 2 and 3 respectively [5].

The control block has two individual synchronized six pulse generators for Thyristor branch 1 and Thyristor branch 2, which are fired at 10°. In this block current is continuously watched and compared with preset value of current. In this paper, we have considered the fault current at 111 A, which is 20% more of nominal value for the concerned branch. Apart from, above we have taken fault current limiting inductor as 10 mH.

This paper has been discussed voltage sag mitigation and suppression of fault current for unsymmetrical/symmetrical faults in a test feeder [6].

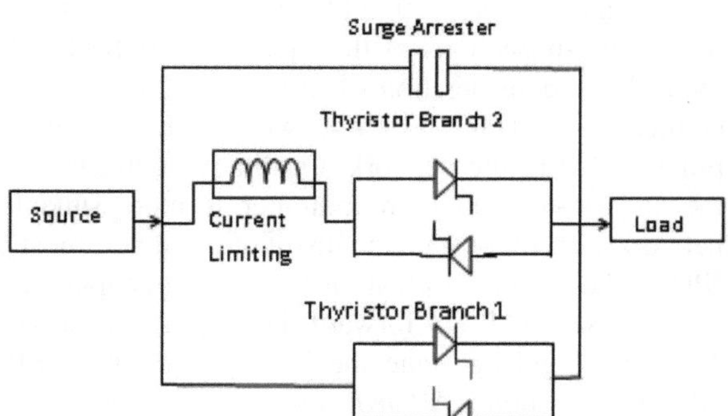

Fig. 1 Basic topology of SSFCL

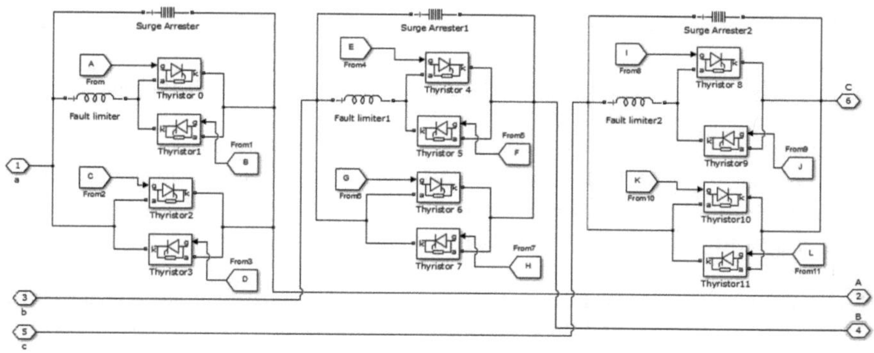

Fig. 2 Simulink model of current limiter block for SSFCL

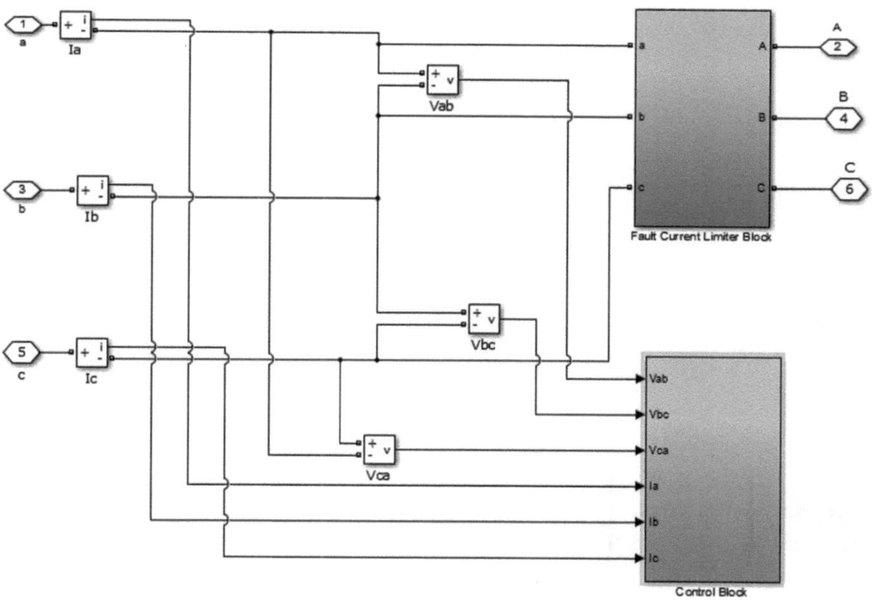

Fig. 3 Simulink model of SSFCL with control block

Test system and SSFCL simulated in a MATLAB SIMULINK environment using system and Simulink parameters as depicted in Table 2. This test system has two parallel feeders as shown in Fig. 4. The F1 and F2 feeder have a constant impedance load, whereas original test system considered F1 feeder load as constant impedance load. Different fault created at load end of F2 feeder, which resulted in reduction of common points of coupling (PCC) and load terminal with the increased transformer current (Fig. 5).

Table 2 System and Simulink parameters

Component	Parameters
Power source	20 kV, $X/R = 5$, 50 Hz, impedance 1.608 Ω, short-circuit level 31.125 MVA
Transformer	10 MVA, 20/6.6 kV, 0.1 PU, *Y*/*y* with ground
Fault Location	Fault at bus 3 and duration 0.02–0.09 s, fault resistance 0.01 Ω
F1 and F2 feeder data	*j*0.314 Ω and *j*0.157 Ω
Load data for F1 and F2 feeder	10 + *j*15.7 Ω and 15 + *j*314 Ω
Fault scenarios	LG fault-phase A to ground, LLG fault-phase A and B with ground, LL fault-phase A and B, LLLG fault-phase A, B and C with ground
Simulink solver parameters	Variable step type with ode23tb, $1e^{-6}$ relative tolerance and $1e^{-1}$ absolute error, simulation time 0.15 s

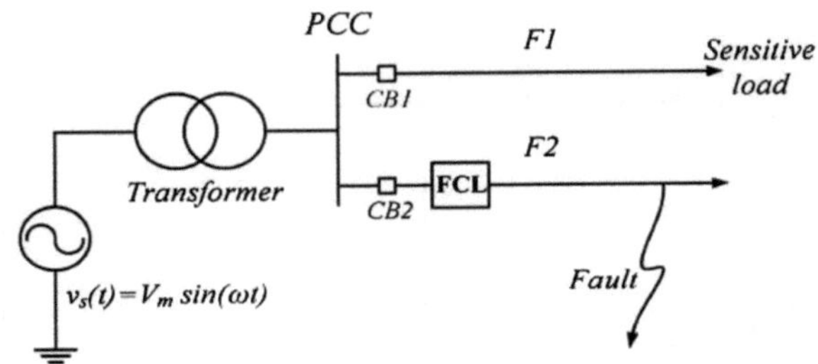

Fig. 4 Single-line diagram of test feeder

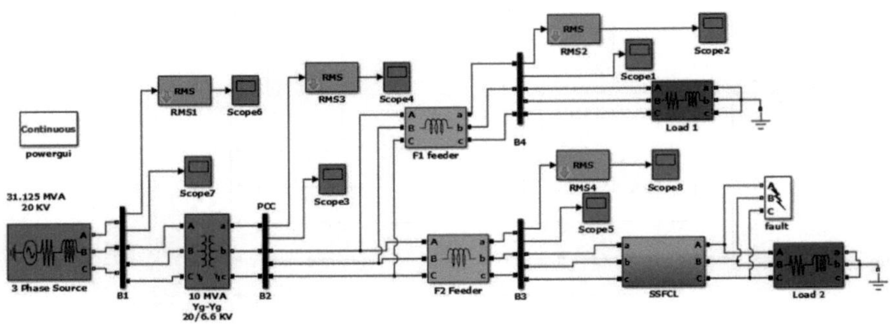

Fig. 5 Simulink model of test system with SSFCL

Results in Tables 3 and 4 shows the impact of SSFCL performance evaluated in voltage mitigation at common points of coupling (PCC)/Bus 4 and fault current reduction for different fault scenarios on bus 3 [6].

As specified in the above Table 4, SSFCL implementation to feeder has a maximum reduction in-line fault found 63% in LL fault, whereas maximum PCC voltage mitigation found 767% in LLG fault. In LLLG fault condition, all phases have same fault current reduction and voltage mitigation. As LLLG fault is most

Table 3 Impact of SSFCL on the voltage profile of PCC and Bus 4 fault under different fault scenarios at Bus 3

Faults	Phase	RMS voltage after fault (KV) at					
		Bus 4		% rise	PCC		% rise
		Without SSFCL	With SSFCL		Without SSFCL	With SSFCL	
LG	A	3.4	4.9	44	3.5	5	43
	B	5.5	5.6	2	5.6	5.6	0
	C	3.4	4.6	35	3.5	4.7	34
LLG	A	0.4	3.8	850	0.45	3.9	767
	B	3.3	4.8	45	3.4	4.9	44
	C	3.2	4.6	44	3.3	4.7	42
LL	A	0.45	3.8	744	0.5	3.9	680
	B	4.9	5.3	8	5	5.4	8
	C	4.8	5.1	6	4.9	5.2	6
LLLG	A	0.45	3.8	744	0.45	3.85	756
	B	0.45	3.8	744	0.45	3.85	756
	C	0.45	3.8	744	0.45	3.85	756

Table 4 Impact of SSFCL on fault current through Bus 3 under different fault scenarios at Bus 3

Faults	Phase	RMS fault current (A)		
		Without SSFCL	With SSFCL	% reduction
LG	A	1800	700	61
	B	93	93	0
	C	93	93	0
LLG	A	1650	730	56
	B	1650	636	61
	C	93	93	0
LL	A	1500	560	63
	B	1460	540	63
	C	93	93	0
LLLG	A	1650	700	58
	B	1650	700	58
	C	1650	700	58

severe and produced greatest electromagnetic stress over associated existing protection system. The variations in PCC voltage and fault current through Bus 3 for a LLLG fault condition without SSFCL are depicted in Fig. 6, whereas impact of SSFCL on PCC voltage and fault current through Bus 3 illustrated in Fig. 7. Thus, it is clear that SSFCL is mitigating voltage and diminishing line fault current significantly.

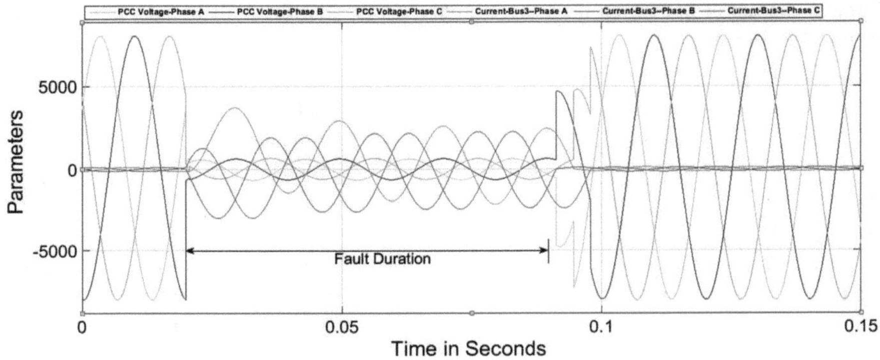

Fig. 6 Parameters of PCC and BUS3 for LLLG without SSFCL

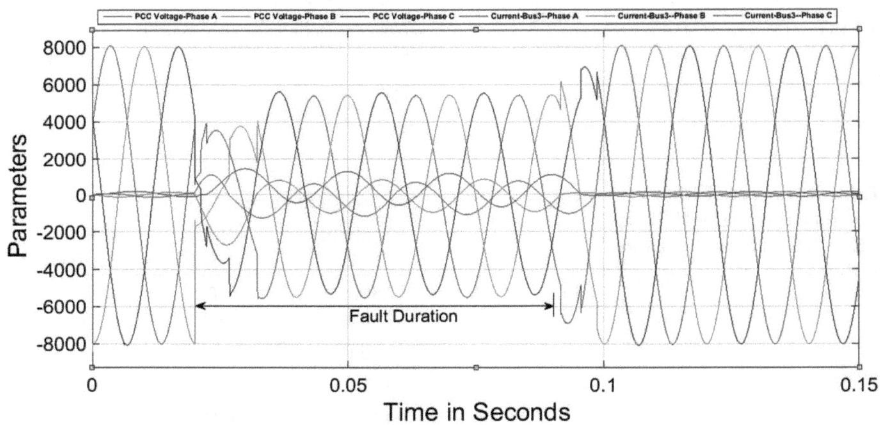

Fig. 7 Parameters of PCC and BUS3 for LLLG with SSFCL

4 Conclusion

Fault current limiting technology is a cost-effective modern fault management system. SSFCL enhanced power quality at PCC bus by voltage sag mitigation and effective fault current control during various fault conditions, which makes DG integration at PCC more reliable. A significant reduction in prospective fault current allows not to upgrade existing protection relay scheme, nor to upgrade existing circuit breakers. As the firing angle gives an approximately sharp decrease in the prospective fault current. Therefore, selection of firing angle depends upon the frequency, severity of the fault. The introduction of fault current limiters is also envisioned post domain R-APDRP and solid-state fault current limiter (SSFCL) deployment in distribution system has an intriguing scope for future studies.

References

1. Central Electrical Authority Reports, Website: www.cea.nic.in/reports/monthly/executivesummary/2016/exe_summary-01.pdf. Retrieved 05 March 2016
2. Restructured Accelerated Power Development and Reforms Programme, Website: www://apdrp.gov.in/What'sNew_Files/BeyondApdrp.pdf. Retrieved 05 March 2016
3. S. Yadav, G.K. Choudhary, R.K. Mandal, Review on fault current limiters. Int. J. Eng. 3(4) (2014)
4. M. Dixit, S. Kulkarni, S. Patil, A. Gupta, P.V. Balasubramanyam, Development of 440 V/800 a high temperature superconducting fault current limiter (SFCL). Indian J. Cryogenics **38**, 1–4 (2013)
5. J.P. Sharma, V. Chauhan, Application of solid state fault current limiter on express feeder for voltage sag mitigation, in IEEE International Conference on Computational Intelligence and Computing Research (ICCIC) (2014). https://dx.doi.org/10.1109/ICCIC.2014.7238449
6. M. Jafari, S.B. Naderi, M.T. Hagh, M. Abapour, S.H. Hosseini, Voltage sag compensation of point of common coupling (PCC) using fault current limiter. IEEE Trans. Power Deliv. **26**(4), 2638–2646 (2011)

Author Index

L. Chandra and A. Dixit (eds.), *Concentrated Solar Thermal Energy Technologies*, Springer Proceedings in Energy, https://doi.org/10.1007/978-981-10-4576-9

MIX
Papier aus verantwortungsvollen Quellen
Paper from responsible sources
FSC® C105338

If you have any concerns about our products,
you can contact us on
ProductSafety@springernature.com

In case Publisher is established outside the EU,
the EU authorized representative is:
Springer Nature Customer Service Center GmbH
Europaplatz 3, 69115 Heidelberg, Germany

Printed by Libri Plureos GmbH
in Hamburg, Germany